Studies in Applied Philosophy, Epistemology and Rational Ethics

Volume 26

About this Series

Studies in Applied Philosophy, Epistemology and Rational Ethics (SAPERE) publishes new developments and advances in all the fields of philosophy, epistemology, and ethics, bringing them together with a cluster of scientific disciplines and technological outcomes: from computer science to life sciences, from economics, law, and education to engineering, logic, and mathematics, from medicine to physics, human sciences, and politics. It aims at covering all the challenging philosophical and ethical themes of contemporary society, making them appropriately applicable to contemporary theoretical, methodological, and practical problems, impasses, controversies, and conflicts. The series includes monographs, lecture notes, selected contributions from specialized conferences and workshops as well as selected Ph.D. theses.

More information about this series at http://www.springer.com/series/10087

Luís Moniz Pereira · Ari Saptawijaya

Programming Machine Ethics

Springer

Luís Moniz Pereira
NOVA Laboratory for Computer Science and Informatics
Faculdade de Ciências e Tecnologia, Universidade Nova de Lisboa
Caparica
Portugal

Ari Saptawijaya
Faculty of Computer Science
Universitas Indonesia
Depok
Indonesia

ISSN 2192-6255 ISSN 2192-6263 (electronic)
Studies in Applied Philosophy, Epistemology and Rational Ethics
ISBN 978-3-319-29353-0 ISBN 978-3-319-29354-7 (eBook)
DOI 10.1007/978-3-319-29354-7

Library of Congress Control Number: 2015960826

Printed on acid-free paper

This springer imprint is published by SpringerNature
The registered company is Springer International Publishing AG Switzerland

For my young granddaughters, Inês and Ana, who will one day come to know ethical machines

Luís Moniz Pereira

For my parents—Yasa and Suli—, Meiriana, and Ananda

Ari Saptawijaya

Foreword

I remember well my first exposure, in 1983, to logic programming (LP). Before this encounter, I had naïvely assumed that first-order logic's computational role was the sterile and static one of providing formulae whose relations capture the inputs and outputs for Turing-level functions (as in the concept of *representability* central to Gödelian incompleteness results), as well as, perhaps (when used in AI), the role of enabling the representation of declarative information, over which reasoning of any type (deductive, and not necessarily resolution; analogical; inductive; etc.) could then occur. But now suddenly in front of me was a remarkably precise, efficacious paradigm that had elevated resolution to a cornerstone, replete with a programming language all its own, glowing with real-world power. Little did I know then, in graduate school, that one of the original founders of the LP paradigm itself, Luís Moniz Pereira, would see to it that three decades later I would receive another jolt of the mind. The second jolt occured when I met Dr. Pereira (for the first time) in Vienna at an OFAI symposium organized by Robert Trappl on machine ethics: There Luís was showing quite concretely in his presentation that LP's power includes its ability to serve as the basis for—as he and co-author Saptawijaya put it in their accurate and ambitious title—programming machine ethics. Of course, the power in question only manifests itself in the hands of wise, creative humans, and much wisdom and creativity are on display in this book, which is my honor to heartily recommend. I know of no other book in which both fundamental principles and finely tuned computational techniques are brought together to yield a self-contained and far-reaching research program. From an unabashed affirmation of functionalism in the philosophy of mind, and contractualism in ethics, to machine ethics, engineering made possible by such powerful techniques as tabling, the book delivers to the world what is currently the most sophisticated, promising, robust approach to machine ethics, period.

Among the book's myriad specific virtues are four to which I draw your attention; the first three: great progress on the counterfactual front; expansion of machine ethics from the traditional case of single agents and their obligations to multiple, interacting agents and their moral dimension; and a very nice assortment of "reading paths," helpfully enumerated in the Preface, which can be taken through the book. Pereira and Saptawijaya are among the first machine ethicists to see the centrality of counterfactuals to this field, and to then make progress on the *programming* of counterfactuals in machine ethics. As to the multi-agent case, this is one of the most exciting aspects of the book. Regarding paths, the reader could for example quickly take two that, together, deliver both of the aforementioned pleasant surprises that I had the good fortune of experiencing: the path of a general introduction to machine ethics from an LP point of view, plus the path of an introduction to some state-of-the-art LP techniques.

And the fourth virtue? You will not find it explicitly presented in the pages of this remarkable book. But you will find it nonetheless. The fourth virtue is humility. Despite leading the way in machine ethics, Pereira and Saptawijaya are steadfastly no-nonsense and content focused, and when they point out that some part of their work is novel, they do so in a matter-of-fact manner that reflects an interest in advancing machine ethics, not two machine ethicists. Their highest priority is progress in machine ethics, pure and simple, and on that commendable metric they achieve nonpareil success.

Troy, NY, USA
December 2015

Selmer Bringsjord
Rensselaer Polytechnic Institute

Preface

Scope

In recent years, systems or agents have become ever more sophisticated, autonomous, and act in groups, amidst populations of other agents, including humans. Autonomous robots or agents have been actively developed to be involved in a wide range of fields, where more complex issues concerning responsibility are in increased demand of proper consideration, in particular when the agents face situations involving choices on moral or ethical dimensions. In medical or elder care, a robot may be confronted with conflicts between attaining its duties to treat the patient and respecting the patient's decision, e.g., in the case where the patient rejects a critical treatment the robot recommends. In the military, where robots from different makers, with diverse purposes, shapes, and sizes have been built and even deployed in wars, will naturally face moral dilemmas, e.g., whether it is permissible for a drone to fire on a house where a target is known to be hiding, but at the same time it is one that also sheltering civilians. In fact, there has been much attention given recently concerning the ethical issue of such autonomous military robots. More recently, a multidisciplinary team of researchers received funding from the U.S. Navy to explore the challenges of providing autonomous agents with ethics.

As these demands become ever more pervasive and ubiquitous, the requirement that agents should function in an ethically responsible manner is becoming a pressing concern. Accordingly, *machine ethics*, which is also known under a variety of names, such as machine morality, computational morality, artificial morality, and computational ethics, emerges as a burgeoning field of inquiry to attend to that need, by imbuing autonomous agents with the capacity for moral decision-making. Clearly, machine ethics brings together perspectives from various fields, among them: philosophy, psychology, anthropology, evolutionary biology, and artificial intelligence. The overall result of this interdisciplinary research is therefore not just important for equipping agents with some capacity for making moral decisions, but also to help better understand morality, via the creation and testing of computational models of ethical theories.

The field is increasingly more so recognized, and its importance has been emphasized in dedicated scientific meetings, book publications, as well as a heightened public awareness to its importance. The Future of Life Institute explicitly identifies machine ethics as one important research priorities in promoting artificial intelligence (AI) research, which is not only capable, but also robust and beneficial. More recently, this initiative, which is supported by top AI researchers from industry and academia, received a significant amount of funding to run a global research program aiming at those priorities. The topic continues to receive wide attention in a variety of conferences, both soliciting papers and promoting panels.

This book reports our investigations on programming machine ethics in two different realms: the individual and collective realms. In studies of human morality, these distinct interconnected realms are evinced too: one stressing above all individual cognition, deliberation, and behavior; the other stressing collective morals, and how they emerged. Of course, the two realms are necessarily intertwined, for cognizant individuals form the populations, and the twain evolved jointly to cohere into collective norms, and into individual interaction.

Content

The main content of the book starts with the groundbreaking work of Alan Turing, in Chap. 1, to justify our *functionalism* stance regarding the modeling of morality, and thus programming machine ethics. In the realm of the individual, computation is vehicle for the study of morality, namely in its modeling of the dynamics of knowledge and cognition of agents. We begin, in Chap. 2, with a survey of research in the individual realm of machine ethics. It provides the context and the motivation for our investigations in addressing moral facets such as permissibility and the dual process of moral judgments by framing together various *logic programming* (LP) knowledge representation and reasoning features that are essential to moral agency, viz., abduction with integrity constraints, preferences over abductive scenarios, probabilistic reasoning, counterfactuals, and updating. Computation over combinations of these features has become our vehicle for modeling the dynamics of moral cognition within a single agent. The investigation of the individual realm of machine ethics reported in this book is a part of the recent Ph.D. thesis of the second author, supervised by the first author, which explores and exploits, for the first time, the aforementioned LP-based knowledge representation and reasoning features for programming machine ethics.

This investigation, elaborated on in Chaps. 3–8 of this book, does not merely address the appropriateness of LP-based reasoning to machine ethics abstractly, but also provides an implementation technique as a basis and testing ground for experimentation of said moral facets. That is, it serves as proof of concept that our understanding of the considered moral facets can in part be computationally modeled and implemented, testing them through a variety of classic moral examples taken off-the-shelf from the morality literature, together with their reported empirical results about their judgments serving as validatory reference.

- Chapter 3 reports on our study of the literature in moral philosophy and psychology for choosing those moral facets and their conceptual viewpoints that are close to LP-based representation and reasoning. We discuss in particular: (1) *moral permissibility*, taking into account the doctrines of double effect and triple effect, and Scanlonian contractualism; (2) *the dual process model* that stresses the interaction between deliberative and reactive processes in delivering moral decisions; and (3) the role of *counterfactual* thinking in moral reasoning.
- Chapter 4 starts with necessary background in logic programming. The reader who is familiar with semantics of logic programs (particularly the stable model and the well-founded semantics), may skip this technical background (in Sect. 4.1), as we simply need them to justify how the results in the examples are obtained. Other sections in this chapter briefly overview the considered LP reasoning features:
  - *Abduction*, whose roles are of scenario generation and of hypothetical reasoning, including the consideration of counterfactual scenarios about the past;
  - *Preferences*, which are enacted for preferring scenarios obtained by abduction;
  - *Probabilistic LP*, which allows abduction to take scenario uncertainty measures into account;
  - *LP updating*, which enables updating the knowledge of an agent, either actual or hypothetical. The latter is used for back-in-time temporary causal interventions specified by counterfactuals in Chap. 6;
  - *LP counterfactuals* permit hypothesizing into the past, and even taking into account present knowledge when so doing;
  - *Tabling* affords solutions reuse (rather than recomputing them), and is employed in joint combination with abduction and updating, as discussed in Chap. 5. Therefore tabling also enables reactivity by picking up readymade solutions, including priorly abduced ones, and those that result from subsequent incremental updating.

In Chap. 4, we also discuss the appropriateness of the above features for representing and reasoning about diverse issues of moral facets tackled in this book.

- Chapter 5 details novel approaches for employing tabling in abduction and updating, viz., tabling abductive solutions in *contextual abduction* (TABDUAL) and the *incremental tabling* of fluents (EVOLP/R), respectively. Tabling in contextual abduction allows the reuse of priorly obtained abduction results in a different context, whereas the use of incremental tabling in updating permits bottom-up propagation of updates, and hence avoids top-down frame axiom computations at a high level. The combined use of tabling in contextual abduction and updating, discussed in Chap. 7, therefore permits the interaction

between the deliberative and reactive processes foreseen by the dual process model.
- Chapter 6 elaborates our LP-based counterfactuals evaluation procedure, concentrating on pure non-probabilistic counterfactual reasoning by resorting to abduction and updating, in order to determine the logical validity of counterfactuals.
- Chapter 7 discusses the three LP systems (ACORDA, PROBABILISTIC EPA, and QUALM), emphasizing how each of them distinctively incorporates a combination of the LP-based representation and reasoning features discussed in Chap. 4.
- Chapter 8 details the applications of ACORDA, PROBABILISTIC EPA, and QUALM for modeling various issues relevant to the chosen moral facets.

The other realm of machine ethics, viz., the collective one, concerns itself with computational moral emergence. The mechanisms of emergence and evolution of cooperation in populations of abstract individuals, with diverse behavioral strategies in co-presence, have been undergoing mathematical study via *evolutionary game theory* (EGT), inspired in part on evolutionary psychology. Their systematic study resorts to simulation techniques, thus enabling the study of aforesaid mechanisms under a variety of conditions, parameters, and alternative virtual evolutionary games. The theoretical and experimental results have continually been surprising, rewarding, and promising.

In the collective realm, the computational study of norms and moral emergence via EGT is typically conducted in populations of rather simple-minded agents. That is, these agents are not equipped with any cognitive capability, and thus simply act from a predetermined set of actions. Our research (of the first author and other co-authors) has shown that the introduction of cognitive capabilities, such as intention recognition, commitment, revenge, apology, and forgiveness, reinforce the emergence of cooperation in the population, comparatively to the absence of such cognitive abilities.

We discuss, in Chap. 9 of this book, how modeling the aforesaid cognitive capabilities in individuals within a networked population shall allow them to fine tune game strategies, and in turn may lead to the evolution of high levels of cooperation. Such driving strategies are associated with moral "emotions" that motivate moral discernment and substantiate ethical norms, leading to improved general conviviality on occasion, or not. To wit, we can model moral agency without explicitly representing embodied emotions, as we know them. Rather, such software-instantiated "emotions" are modeled as (un)conscious strategic heuristics empowered in complex evolutionary games. Moreover, modeling such capabilities in individuals within a population may help us understand the emergent behavior of ethical agents in groups, in order to implement them not just in a simulation, but also in the real world of future robots and their swarms.

The EGT techniques mentioned in Chap. 9 are rather standard and hence use rather standard EGT software, whose originality of use relies, as usual, on infusing it with the specific equations of our analytical mathematical models, plus the parameters for the attending simulations. They can be followed up in detail in our own references therein, many of them in open-access publications. Because they are substantially mathematically sophisticated and extensive in nature, we thought best to leave them out of the book. Suffice it to say that these EGT techniques enable us to obtain our results, regarding mixed populations of individuals and their strategies, either analytically, by simulation, or both.

Having contemplated the two distinct realms of machine ethics, a fundamental question then arises, concerning the study of individual cognition in small groups of frequently morally interacting multi-agents that can choose to defect or cooperate with others. That is, whether from such a study we can obtain results equally applicable to the evolution of populations of such agents; and vice versa, whether the results obtained in the study of populations carry over to groups of frequently interacting multi-agents, and under what conditions. This issue is discussed in Chap. 10, by bringing to the fore congenial present views and research on the evolution of human morality, hoping to reinforce the bridging ideas and computational paradigms we set forth. Moreover, we take for granted that computational and robotic models can actually provide abstract and concrete insight on emerged human moral reality, irrespective of the distinct embodiments of man and machine.

Reading Paths

The book is best read sequentially, chapter by chapter. Nevertheless, several alternative reading paths are possible, as shown below. These reading paths can also be first read cursively, by simply reading the introductions and concluding remarks of each of the corresponding chapters. Those in parentheses provide some necessary background, technical details, and further references for their respective topic. They may safely be skipped and read only if needed. Moreover, the explanations set forth in the previous section, about the motivation and resulting use of the main techniques, may well suffice for the general reader who is not concerned with the details of how they are brought about and integrated, and is willing to believe that our implemented techniques actually work.

- A general survey of approaches to machine ethics:

$$1 \rightarrow 2 \rightarrow 3 \rightarrow (4.1) \rightarrow 4.2 \text{ till } 4.8 \rightarrow 10 \rightarrow 11$$

- The individual realm of machine ethics via logic programming (LP):

$$1 \rightarrow 2 \rightarrow 3 \rightarrow 4 \rightarrow 5 \rightarrow 6 \rightarrow 7 \rightarrow 8 \rightarrow 11$$

- Focusing on abduction and preferences over abductive scenarios:

$$(3) \rightarrow 4.1 \rightarrow 4.2 \rightarrow 4.3 \rightarrow 7.1 \rightarrow 8.1$$

- Focusing on abduction and probabilistic LP:

$$(3) \rightarrow 4.1 \rightarrow 4.2 \rightarrow 4.4 \rightarrow 7.2 \rightarrow 8.2$$

- Focusing on abduction, updating, and tabling:

$$(3) \rightarrow 4.1 \rightarrow 4.2 \rightarrow 4.5 \rightarrow 4.7 \rightarrow 5 \rightarrow 7.3 \rightarrow 8.3$$

- Focusing on counterfactuals:

$$(3) \rightarrow 4.1 \rightarrow 4.2 \rightarrow 4.5 \rightarrow 4.6 \rightarrow 6 \rightarrow (7.3) \rightarrow 8.3.2$$

- The collective realm of machine ethics via evolutionary game theory:

$$1 \rightarrow 9 \rightarrow 10 \rightarrow 11$$

- General LP engineering techniques:
  - Abduction with tabling: $4.1 \rightarrow (4.2) \rightarrow (4.7) \rightarrow 5.1$
  - Updating with tabling: $4.1 \rightarrow (4.5) \rightarrow (4.7) \rightarrow 5.2$
  - Counterfactuals: $4.1 \rightarrow (4.6) \rightarrow 6$

Audience

The primary audience for this book are the researchers and postgraduate students in the cognitive sciences, artificial intelligence, robotics, philosophy, and ethics. The secondary audience for this book, taking into account the above reading suggestions, are the curious academic and general publics; undergraduates looking for topics of research; science journalists; science and society forums; legislators and the military concerned with machine ethics.

Lisbon, Portugal — Luís Moniz Pereira
Jakarta, Indonesia — Ari Saptawijaya

Acknowledgments

Profound thanks are due to our co-authors of joint published work cited in Chap. 9, without which the personal summing up and specific philosophical viewpoints, in Chap. 9 and in its subsequent chapter, would not have been possible at all. Specifically, we thank the co-authors of joint papers, The Anh Han, Luis Martinez-Vaquero, Francisco C. Santos, and Tom Lenaerts, for use of summaries of material from diverse joint publications detailed in the references.

We thank David S. Warren and Theresa Swift for their expertise in helping us with XSB Prolog, as well as for their specific suggestions in the development of tabling in abduction and updating; these are two essential LP-based knowledge representation and reasoning features employed in the approaches and applications in this book. We thank Robert Kowalski for the discussion on the idea of contextual abduction and rule name fluent mechanism LP updating. We also acknowledge Emmanuelle Dietz for our discussion on counterfactuals, Gonçalo Lopes for discussing the use of ACORDA as shown in this book, and Miguel Calejo for a number of suggestions to improve QUALM in future.

Furthermore we thank João Alexandre Leite, Alexandre Miguel Pinto, Fariba Sadri, and Theresa Swift, the jury members of the second author's Ph.D. thesis, supervised by the first author, whose suggestions have improved the book, as it being an outgrowth of core parts of the thesis, conjoined with distinct and diverse subsequent substantial work.

Luís Moniz Pereira acknowledges ongoing support from Fundação para a Ciência e a Tecnologia (FCT/MEC) in Portugal, via the NOVA LINCS grant PEst UID/CEC/04516/2013. Ari Saptawijaya also acknowledges the support from FCT/MEC that has funded his doctoral study through grant SFRH/BD/72795/2010, the NOVA LINCS research center and the Departamento de Informática, Universidade Nova de Lisboa, for the supplementary funding to attend scientific conferences during his study, and the Faculty of Computer Science, Universitas Indonesia, for permitting him to take a leave of absence to pursue his doctoral study.

Last but not least, we heartily thank Prof. Selmer Bringsjord for his ready acceptance, at short notice, to write the Foreword to this book.

Contents

Chapter 1
Turing, Functionalism, and Emergence

The main content of this chapter addresses the relevance of the ground breaking work of Alan Turing to justify our functionalism stance regarding the modeling of morality. Building intelligent machines seeks a partial understanding of the emergence of higher-level properties, like morality. Functionalism holds that the material substrate is not of the essence, and that it suffices to realize equivalent functionality albeit by way of a different material vehicle. The most fruitful inquires into the nature of "mind" or "general intelligence" shall include the use of Artificial Intelligence to simulate complex mental operations.

1.1 Turing Is Among Us

Turing's relevance arises from the timelessness of the issues he tackled, and the innovative light he shed upon them [8]. He first defined the algorithmic limits of computability, via an *effective* well-specified mechanism, and showed the generality of his definition by proving its equivalence to other general, but less algorithmic and non-mechanical, more abstract formulations of computability. His originality lies on the essential simplicity of the mechanism invoked—the now dubbed Turing Machines (or programs), which he called A-Machines—and the proof of existence of a Universal A-Machine (i.e., the digital computer, known in academia as the Universal Turing Machine), which can simulate any other A-Machine, that is, execute any program.

Interestingly, he raised the issue of whether human beings are a measure for his "machines", and, in mechanizing human cognition, Turing implicitly introduced the modern perspective since known as "functionalism". According to this paradigm, what counts is the realization of function, independently of the hardware embodying it. Such "multiple realization" is afforded by the very simplicity of his devised mechanism, relying solely on the manipulation of discrete information, where data and instructions are both represented just with symbols. The twain are stored in memory,

L.M. Pereira and A. Saptawijaya, *Programming Machine Ethics*,
Studies in Applied Philosophy, Epistemology and Rational Ethics 26,
DOI 10.1007/978-3-319-29354-7_1

instructions doubling as data and as rules for acting—the stored program. To this day, no one has invented a computational mechanical process with such general properties, which cannot be theoretically approximated with arbitrary precision by some A-Machine, where any interactions with the world outside are captured by Turing's innovative concept and definition of "oracle"—the very word employed by him for the purpose—, as a means to interrogate that world by posing queries to one or more outside oracles. This concept of oracle is regularly taught in computer science today, namely in the essential study of computation complexity, though not every student knows it came from Turing. In the midst of a computation a query may be posed to an outside oracle about the satisfaction of some truth, and the computation continued once an answer obtained, rather than the computer testing for an answer in a possibly infinite set of them.

Turing further claimed that his machines could simulate the effect of *any* activity of the mind, not just a mind engaged upon a "definite method of proceeding" or algorithm. He was clear that discrete state machines included those with learning or self-organizing abilities, and stressed that these still fall within the scope of the computable. Turing drew attention to the apparent conflict between self-organization and the definition of A-Machines as having fixed tables of behavior, but sketched a proof that self-modifying machines are still definable by an unchanged instruction set [5, 7]. The promise of this approach in studies of morality is that it represents a universal functionalism, the terms of which enable the bringing together of the ghosts in the several embodied machines (silicon-based, biological, extra-terrestrial or otherwise), to promote their symbiotic epistemic co-evolution, as they undertake moral action within a common moral theater.

1.2 Functionalism

The principle of the distinction between software and hardware appears clear-cut with the advent of the digital computer and its conceptual precursor, the Universal Turing Machine. The diversity of technologies employed to achieve the same function, confirms it ever since the first computers. One program is executable in physically different machines, precisely because the details of its execution below an ascertainable level of analysis are irrelevant, as long as an identical result at the level of discourse is produced. That said, however, the distinction between hardware and software is not so clear as it might seem. Hardware is not necessarily represented by physical things but rather by what, at some level of analysis, is considered fixed, given, and whose analysis or non-analyzability is irrelevant for the purpose at hand. Historically, in the first computers, that level coincided with that of the physical parts of the machine. Subsequently, especially due to rapidly increasing computing power, "hardware" has become increasingly "soft", with the physical basis for the hardware/software distinction finally blurred by the concept of the "abstract machine": a fixed collection of mathematically defined instructions supporting a set of software functions, independently of the particular physical processes underlying the implementation of the abstract machine, that is, realizing it.

Hence, "multiple realization" stands for the thesis that a mental state can be "realized" or "implemented" by different physical states. Beings with different physical constitutions can thus be in the same mental state, and from these common grounds can cooperate, acting in mutual support (or not). According to classical functionalism, multiple realization implies that psychology is autonomous: in other words, biological facts about the brain are irrelevant [1]. Whether physical descriptions of the events subsumed by psychological generalizations have anything in common is irrelevant to the truth of the generalizations, to their interestingness, to their degree of confirmation, or, indeed, to any of their epistemological important properties [4].

Functionalism has continued to flourish, being developed into numerous versions by thinkers as diverse as David Marr, Daniel Dennett, Jerry Fodor, and David Lewis [3, 4]. It helped lay the foundations for modern cognitive science, being the dominant theory of mind in philosophy today. In the latter part of the 20th and early 21st centuries, functionalism stood as the dominant theory of mental states. It takes mental states out of the realm of the "private" or subjective, and gives them status as entities open to scientific investigation. Functionalism's characterization of mental states in terms of their roles in the production of behavior grants them the causal efficacy that common sense takes them to have. In permitting mental states to be multiply realized, functionalism offers an account of mental states compatible with materialism, without limiting the class of minds to creatures with brains like ours [6].

1.3 Emergence

Biological evolution is characterized by a set of highly braided processes, which produce a kind of extraordinarily complex combinatorial innovation. A generic term frequently used to describe this vast category of spontaneous, and weakly predictable, order-generating processes, is "emergence". This term became a sort of signal to refer to the paradigms of research sensitive to systemic factors. Complex dynamic systems can spontaneously assume patterns of ordered behaviors not previously imaginable from the properties of their constitutive elements or from their interaction patterns. There is unpredictability in self-organizing phenomena—preferably called "evolutionary" [9]—with considerably variable levels of complexity, where "complexity" refers to the emergence of collective properties in systems with many interdependent components. These components can be atoms or macromolecules in physical or biological contexts, and people, machines or organizations in socioeconomic contexts.

What does emerge? The answer is not something defined physically but rather something like a shape, pattern, or function. The concept of emergence is applicable to phenomena in which the relational properties predominate over the properties of the compositional elements in the determination of the ensemble's characteristics.

Emergence processes are due to starting configurations and interaction topologies, not intrinsic to the components themselves [2]. This functionalism is, almost by definition, anti substance-essence, anti vital-principle, anti monopoly of *qualia*.

1.4 Concluding Remarks

Building intelligent machines may seek a partial understanding of the emergence of higher-level properties, like morality. Here, functionalism affirms the salience of the results of this work in assessing, for example, human morality. Again, functionalism holds that the material substrate is not of the essence, and that it suffices to realize equivalent functionality albeit by way of a different material vehicle. Moreover, distinct roads to the same behavior may be had, thereby adding to our understanding of what, say, "general intelligence" or "mind" means. Thus, on our estimation, the most fruitful inquires into the nature of "mind" or "general intelligence" will certainly include the use of Artificial Intelligence aided in time by the embryonic field of artificial emotions, qua strategies, to simulate complex mental operations, as already foreseen in [9].

References

1. Boden, M.A.: Information and cognitive science. In: Adriaans, P., van Bentham, J. (eds.) Philosophy of Information. North Holland, Elsevier, Amsterdam (2008)
2. Deacon, T.W.: The hierarchic logic of emergence: untangling the interdependence of evolution and self-organization. In: Weber, H.W., Depew, D.J. (eds.) Evolution and Learning: The Baldwin Effect Reconsidered. The MIT Press, Cambridge (2003)
3. Dennett, D.C.: Sweet Dreams: Philosophical Obstacles to a Science of Consciousness. The MIT Press, Cambridge (2005)
4. Fodor, J.A.: Special sciences, or the disunity of science as a working hypothesis. Synthèse **28**, 77–115 (1974)
5. Hodges, A.: Alan Turing: One of the Great Philosophers. Phoenix, London (1997)
6. Levin, J.: Functionalism. In: Zalta, E.N. (ed.) The Stanford Encyclopedia of Philosophy, Fall 2013 edn. Center for the Study of Language and Information, Stanford University. http://plato.stanford.edu/archives/fall2013/entries/functionalism/ (2013)
7. McDermott, D.: Mind and Mechanism. The MIT Press, Cambridge (2001)
8. Pereira, L.M.: Turing is among us. J. Log. Comput. **22**(6), 1257–1277 (2012)
9. Turing, A.M.: Computing machinery and intelligence. Mind **59**, 433–460 (1950)

Part I
The Individual Realm

Chapter 2
The Individual Realm of Machine Ethics: A Survey

In this chapter, a survey of research in machine ethics is presented, providing the context and the motivation for our investigations. The survey concerns the individual realm of machine ethics, whereas the background to other realm, the collective one, is broached in Chap. 9, namely Sects. 9.1 and 9.2.1. The first realm views computation as a vehicle for representing moral cognition of an agent and its reasoning thereof, which motivates our investigation for employing Logic Programming (LP) knowledge representation and reasoning features with respect to the individual realm of machine ethics. On the other hand, the second realm emphasizes the emergence, in a population, of evolutionarily stable moral norms, of fair and just cooperation, that ably discards free riders and deceivers, to the advantage of the whole evolved population. It provides a motivation of our research for introducing cognitive abilities, such as intention recognition, commitment, revenge, apology, and forgiveness, to reinforce the emergence of cooperation in the collective realm of machine ethics.

2.1 Truth-Teller and SIROCCO

Truth- Teller [35] is a system that qualitatively compares a pair of ethical dilemma cases about whether to tell the truth and extracts ethically salient similarities and differences of the reasons for telling the truth (or not), from the perspective of the agent faced with the dilemma. The representation of a case is manually constructed from the interpretation of the story. Semantic networks are employed to represent the truth telling episodes (including the actors involved, their relationships, possible actions and reasons supporting possible actions), a hierarchy of relationships (familial, commercial, etc.), and a hierarchy of reasons for or against telling the truth based on the formulation in [14]. The representation is then analyzed by case-based (casuistic) reasoning in several steps:

L.M. Pereira and A. Saptawijaya, *Programming Machine Ethics*,
Studies in Applied Philosophy, Epistemology and Rational Ethics 26,
DOI 10.1007/978-3-319-29354-7_2

- First, a mapping between the reasons in two cases is built, viz., by matching similar and marking distinct reasons.
- The second step qualifies: (1) the relationships among actors, actions, and reasons in one case; and (2) the mappings of these objects to those in the other considered case based on considerations such as criticalness, actors' roles and alternative actions.
- The third step points out similarities and differences of cases, with respect to predefined comparison contexts, whether the considered reasons apply to both cases, apply more strongly in one case than another, or apply to only one case.

The analysis result is then summarized in a comparison text.

SIROCCO [34] also employs case-based (casuistic) reasoning, but unlike TRUTH- TELLER, it accepts an ethical dilemma case and retrieves ethics principles and past cases that are relevant to the target case. It is developed in order to operationalize general abstract ethics principles, as ethicists often record their explanations of how and why they applied and reconciled principles in resolving specific cases. SIROCCO particularly addresses a domain of engineering ethics, taking into account ethics code of the National Society of Professional Engineers (NSPE) [38] and the cases decided by its Board of Ethical Review (BER). The BER's decision making indicates that several operationalization techniques are applied, which include linking ethics code and past cases with the facts of the considered case, grouping codes and cases so they can be cited in support of a conclusion, and reusing some reasoning applied in past cases to the context of a new case. In SIROCCO, cases are represented in general using the Ethics Transcription Language (ETL) [34] as chronological narratives of facts involving actors, actions participated by actors, and temporal relations between facts. The representation of a source case is particularly extended with its analysis by BER, which captures an operationalization of NSPE ethics codes. In its retrieval process, SIROCCO first computes the best N matches of source cases with respect to the ETL fact matching between the target case and each source case, and subsequently finds a structural mapping using A^* search between the target case and the best N matches [15]. The goal is to map facts of a source to a corresponding fact in the target at the same level of abstraction, while keeping a consistent mapping between their actors and temporal relations. Finally, it analyzes the results of multiple source cases (rather than of a single best match) to generate suggestions for the target case, such as relevant principles and relevant source cases.

2.2 JEREMY and W.D.

JEREMY [6] is a system that follows the theory of act utilitarianism. This theory maintains that an act is morally right if and only if the act maximizes the good, viz., the one with the greatest net good consequences, taking into account all those affected by the action [46]. In JEREMY, hedonistic act utilitarianism is particularly adopted, where the pleasure and displeasure of those affected by each possible action are considered. This is manifested by three components with respect to each affected person p: (1) the intensity I_p of pleasure/displeasure, scaled between -2 and 2; (2) the duration D_p of the pleasure/displeasure, in days; and (3) the probability P_p that this pleasure/displeasure will occur. The total net pleasure for each action a is computed as follows:

$$Total_a = \Sigma_{p \in Person}(I_p \times D_p \times P_p).$$

The right action is the one giving the highest total net pleasure $Total_a$.

In order to respond to critics of act utilitarianism, another prototype, W.D. [6], is developed to avoid a single absolute duty. That is, it follows several duties, where in different cases a duty can be stronger than (and thus overrides) the others, following the theory of prima facie duties of Ross [45]. This theory comprises duties like fidelity (one should honor promises), reparation (one should make amends for wrongs done), gratitude (one should return favors), justice (one should treat people as they deserve to be treated), beneficence (one should act so as to bring about the greatest good), non-maleficence (one should act so as to cause the least harm), and self-improvement (one should develop one's own abilities/talent to the fullest).

In W.D., the strength of each duty is measured by assigning it a weight, capturing the view that a duty may take precedence over another. W.D. computes, for each possible action, the weighted sum of duty satisfaction, and returns the greatest sum as the right action. In order to improve the decision, in the sense of conforming to a consensus of correct ethical behavior, the weight of a duty is allowed to be adjusted through a supervised learning, by acquiring suggested action from the user. This weight adjustment to refine moral decision is inspired by reflective equilibrium of Rawls [41]: reflecting on considered judgments about particular cases and revising any elements of these judgments (principles that govern these judgments, theories that bear on them, etc.) wherever necessary, in order to achieve an acceptable coherence amongst them, the so-called equilibrium [19]. In [6], it is however unclear which supervised learning mechanism is actually implemented in W.D.

2.3 MEDETHEX and ETHEL

The theory of prima facie duties is further considered in [2, 8], while also concretely employing machine learning to refine its decision making. As in W.D., the employing of machine learning is also inspired by reflective equilibrium of Rawls [41], viz., to generalize intuition about particular cases, testing this generalization on further cases, and then repeats this process to further refine the generalization towards the end of developing a decision procedure that agrees with intuition.

The first implementation is MEDETHEX [8], which is based on a more specific theory of prima facie duties, viz., the Principle of Biomedical Ethics of Beauchamp and Childress [13]. The considered cases are a variety of the following type of ethical dilemma [7]:

A healthcare professional has recommended a particular treatment for her competent adult patient, but the patient has rejected it. Should the healthcare professional try to change the patient's mind or accept the patient's decision as final?

The cases thus involve only two possible actions, viz., (1) accepting a patient's decision to reject a treatment; and (2) trying to convince him to change his mind. Furthermore, the cases are constrained to three of four duties in [13], viz., respecting for the autonomy of the patient, not causing harm to the patient (non-maleficence), and promoting patient welfare (beneficence).

MEDETHEX is implemented using Inductive Logic Programming (ILP) [36] to learn the relation $supersedes(A_1, A_2)$, i.e., whether action A_1 supersedes (viz., is ethically preferable to) action A_2. The training (positive) examples comprise cases, where each case is associated with an estimate satisfaction/violation value of each duty for each possible action (scaled from -2 to 2) and the ethically preferred action for the case. The negative examples are obtained by simply exchanging the preferred action from the positive training examples. The relation $supersedes(A_1, A_2)$ is then learned from these positive and negative examples, expressing it in terms of the lower bounds for difference of values of the considered duties between the two actions A_1 and A_2.

Similar to MEDETHEX, ETHEL is also based on the Principle of Biomedical Ethics of Beauchamp and Childress [13], but applied to the domain of eldercare with the main purpose to remind a patient to take his/her medication, taking ethical duties into consideration. It also decides, after a patient has been reminded, whether to accept his/her refusal to take the medication (in which case a further reminder may take place) or to notify an overseer (e.g., a medical doctor) instead.

ETHEL is also implemented using ILP, following a similar technique employed in MEDETHEX to learn the same relation $supersedes(A_1, A_2)$; this relation is also defined in terms of the lower bounds for difference of values of the corresponding duties between actions A_1 and A_2. Unlike MEDETHEX, due to the reminder feature, the satisfaction/violation values of duties for each action in ETHEL are adjusted over time. This adjustment is determined by several factors, such as the maximum amount of harm if the medication is not taken, the number of hours for this maximum harm to occur, etc.; this information is obtained from the overseer. Adjusting the satisfaction/violation values of duties permits ETHEL to remind (or not) the patient to take his/her medication as well as to notify (or not) the overseer at ethically justifiable moment.

ETHEL has been deployed in a robot prototype, capable to find and walk toward a patient who needs to be reminded of medication, to bring the medication to the patient, to engage in a natural-language exchange, and to notify an overseer by email when necessary [3].

There has been work going beyond ETHEL from the same authors, viz., GENETH [4, 5], which is implemented using the same ILP technique.

2.4 A Kantian Machine Proposal

A more philosophical tone of machine ethics is presented in [40], where he argues that rule-based ethical theories like the first formulation of Kant's categorical imperative ("Act only according to that maxim whereby you can at the same time will that it should become a universal law without contradiction" [30]) appear to be promising for computational morality, because of their computational structure for judgment. Three views on how to computationally model categorical imperative are envisaged.

First, in order for a machine to maintain consistency in testing ethical behavior, it should be able to construct a moral theory that renders individual maxims to be universally quantified (over circumstances, purposes, and agents) and to map them onto deontic categories, viz., forbidden, permissible, and obligatory action. Deontic logic is regarded as an appropriate formalism with respect to this first view. He abstractly refers to schemata for the three deontic categories, that for every agent, circumstance C, and purpose P:

- Action A is obligatory: $(C \text{ and } P) \rightarrow A$.
- Action A is forbidden: $(C \text{ and } P) \rightarrow \neg A$.
- Action A is permissible: $\neg((C \text{ and } P) \rightarrow A)$ and $\neg((C \text{ and } P) \rightarrow \neg A)$.

where a candidate maxim should be an instance of these three schemata.

Powers suggests that mere consistency is not sufficient for a maxim. Instead, its consistency should also be checked with other existing facts or background theory. This leads to his second view, viz., the need of common-sense reasoning in the categorical imperative to deal with contradiction. For this view, he refers to non-monotonic logic, which is appropriate to capture defeating conditions to a maxim. In this regard, he particularly resorts to default logic of Reiter [43] as a suitable formalism, that adding the default rules allows maxims to contradict the background set of facts and common-sense rules without introducing inconsistency.

In his third view, Powers contemplates on the construction of a coherent system of maxims, where he sees such construction analogous to the belief revision problems. In the context of bottom-up construction, he envisages an update procedure for a machine to update its system of maxims with another maxim, though it is unclear to him how such an update can be accomplished.

The formalisms in these three views are only considered abstractly and no implementation is referred to address them.

2.5 Machine Ethics via Theorem Proving

In [16], mechanized multi-agent deontic logic is employed with the view that ethically correct robot behaviors are those that can be proved in a deontic logic. For obtaining such a proof of ethically permissible actions, they resort to a sequent-based natural-deduction of Murakami [37] axiomatization of Horty's utilitarian formulation of multi-agent deontic logic [26]. This deduction system is encoded in the interactive theorem prover ATHENA [9]. The use of *interactive* theorem prover is motivated by the idea that an agent operates according to ethical codes bestowed on them, and when its automated reasoning fails, it suspends its operation and asks human guidance to resolve the issue.

Taking an example in health care, where two agents are in charge of two patients with different needs (patient H_1 depends on life support, whereas patient H_2 on very costly pain medication), two actions are considered: (1) terminate H_1's life support to secure his organ for five humans; and (2) delay delivery of medication to H_2 to conserve hospital resources. The approach in [16] begins with supposing several candidates of ethical codes, from harsh utilitarian (that both terminates H_1's life and delay H_2 medication) to most benevolent (neither terminates H_1's life nor delay H_2 medication); these ethical codes are formalized using the aforementioned deontic logic. The logic additionally formalizes behaviors of agents and their respective moral outcomes. Given these formalizations, ATHENA is employed to query each ethical code candidate in order to decide which amongst them should be operative, meaning that the best moral outcome (viz., that resulting from neither terminates H_1's life nor delay H_2 medication) is provable from the operative one.

2.6 Particularism versus Generalism

A computational model to study the dispute between particularism and generalism, is explored in [23, 25]. Moral generalism stresses the importance of moral principles and their *general* application in moral reasoning, whereas moral particularism favors on the view that moral reasoning (and decisions) depend on cases and not on a general application of moral principles to cases [18].

In [23], different ethical principles (of Aristotle, Kant, and Benjamin Constant) are modeled using answer set programming, implemented with ANSPROLOG* [11]. The aim is to show that non-monotonic logic is appropriate to address the opposition between generalism and particularism by capturing justified exceptions in general ethics rules. The tension between these two viewpoints is exemplified by a classic dilemma about lying: in a war situation one hides a friend who is wanted by the military force, raising to a dilemma whether he should tell the truth, denouncing his friend to the military, which leads to the murder of his friend.

In order to model this dilemma in the view of Aristotle's ethics (viz., choosing the least unjust action), several possible actions are conceived, e.g., tell the truth, tell a lie, etc., and facts about consequences of these actions are defined. Predicate $unjust(A)$ is then defined by assessing whether the consequence of A is worse than the consequence of other actions, via predicate $worse/2$, whose parameters are the consequences of two considered actions. Depending on the definition of $worse/2$, the answer sets may be split into one part corresponding to telling the truth and the other part to telling a lie. The model itself does not provide a mechanism to prefer among these answer sets, though it illustrates that an ad hoc preference is possible by explicitly changing the definition of $worse/2$ predicate so as all answer sets contain the action of telling a lie (providing that murder has worse consequence than that of all other actions).

Ganascia [23] also contrasts Kant's categorical imperative and Constant's objection. For Kant's categorical imperative, a rule such as:

$$act(P, A) \leftarrow person(P), action(A), act(\text{“}I\text{”}, A)$$

is defined to universalize a maxim: it stipulates that if "I" act in such A, all person (P) could act the same. This view does not require preferences among different actions, but emphasizes possible consequences of a maxim that cannot be universalized, e.g., a society where nobody can be trusted: $untrust(P) \leftarrow act(P, tell(P, lie))$. To this end, while lying can be admitted in an answer set, the answer set reflects a world where nobody can be trusted.

While this Kantian view aims at upholding generality of ethics principles, Constant's theory authorizes principles that tolerate exceptions. The lying dilemma is modeled by capturing a more specific principle for telling the truth: we always have to tell the truth, except to someone who does not deserve it. This is achieved in the model by: (1) not only considering the transmitter of the speech (as in the Kantian model), but also the receiver; and (2) using default negation to express the principle in a way that one should always tell the truth, except when the receiver is a murderer.

In [25], the dispute between particularism and generalism is addressed using artificial neural networks. More specifically, simple recurrent networks are trained with cases about permissibility of actions involving killing and allowing to die.

The input for a network encodes the actor, the recipient of the action and the motive or the consequence of the action (e.g., killing in self-defence, allowing one to die to save many innocents, etc.), but without the provision of explicit moral rules, whereas the output of the network determines the permissibility of an input case. The experiments are performed on several networks that are trained differently. For instance, one network is trained by classifying permissibility based on the motive or the consequence of the action (irrespective whether the action is killing or allowing to die), whereas another network is trained by distinguishing the action killing from allowing to die.

By employing these trained networks to classify test cases, one result suggests that acting in self-defence contributes to permissibility, whereas actions that lead to the deaths of innocents are impermissible. Further analysis on the similarity of hidden unit activation vector between cases suggests that killing and allowing to die are making different contributions to the similarity spaces for different trained networks. Nonetheless, the networks admittedly learn some principles in general, though it cannot be directly expressed in classical, discrete representational structure. The experiments thus show that the behavior of the networks is in agreement with the so-called *contributory standards* of moral principles [18]—a middle ground between particularist and generalist—which allows more than one principle to be applicable to a case, as each specifies how things are, only in a certain respect.

2.7 Concluding Remarks

In this chapter the open and broad nature of research in machine ethics has been illustrated. On the one hand, it spans a variety of morality viewpoints, e.g., utilitarianism, Kant's categorical imperative, prima facie duties, particularism and generalism. On the other hand, a number of approaches have been proposed to model these viewpoints, with some assumptions to simplify the problem, which is somewhat unavoidable, given the complexity of human moral reasoning. The open nature of this research field is also indicated by different purposes these approaches are designed for (e.g., retrieving similar moral cases, explicitly making moral decisions, or finding operative moral principles).

TRUTH- TELLER and SIROCCO point out the role of knowledge representation (using semantic networks and its specific language ETL, respectively) to represent moral cases in a sufficiently fine level of detail, and rely on such representation for comparing cases and retrieving other similar cases. This form of representation is not so emphasized in JEREMY and W.D., as they reduce the utilitarian principle and duties into numerical values within some scale. Unlike TRUTH- TELLER and SIROCCO, JEREMY and W.D. aim at explicitly making moral decisions; these decisions are determined by these values through some procedure capturing the moral principles followed. Such quantitative valuation of duties also forms the basis of MEDETHEX and ETHEL, though some basic LP representation is employed for representing this valuation in positive and negative instances (which are needed for their ILP learning mechanism), as well as for representing the learned principle in the form of a LP rule. This learned principle, in terms of these numerical values, determines moral decisions made by these systems.

The employment of logic-based formalisms in the field, notably deontic logic, to formalize moral theory appears in the Kantian machine proposal of Powers [40]. Indeed, the insufficient of abstract logic-based formalism for rule-based ethical theories is identified in this proposal, emphasizing the need of non-monotonic reasoning in order to capture defeating conditions to a maxim. Moreover, the proposal also points out the importance of an update procedure to anticipate updating a system of maxims with another. Unfortunately there is no concrete realization of this proposal. In [16], an interactive theorem prover is employed to encode a specific deontic logic formalism in order to find an operative moral principle (amongst other available ones) in the form of proof. The use of theorem prover in this approach however does not concern the non-monotonic reasoning and the moral updating issues raised by Powers [40].

The issue of non-monotonic reasoning becomes more apparent in the study about particularism versus generalism. Ganascia [23] demonstrates in a concrete moral case how non-monotonic reasoning can be addressed in LP—more specifically in answer set programming—using defeasible rules and default negation to express principles that tolerate exception. From a different perspective, the experiments with artificial neural networks [25] also reveal that more than one principle may be applicable to similar cases that differ in a certain aspect (e.g., motives, consequences, etc.), thus upholding morality viewpoints that tolerate exceptions.

While the survey in this chapter shows that several logic-based approaches have been employed in machine ethics, the use of LP has not been much explored in the field despite its potential:

- Like TRUTH- TELLER and SIROCCO, LP permits declarative knowledge representation of moral cases with sufficiently level of detail to distinguish one case from other similar cases. Indeed, except the philosophical approach by Powers [40], all other approaches anchor the experiments to concrete moral cases, indicating that representing moral principles alone is not enough, but the principles need to be materialized into concrete examples. Clearly, the expressivity of LP may extend beyond basic representation of (positive/negative) example facts demonstrated in MEDETHEX and ETHEL.
- Given its declarative representation of moral cases, appropriate LP-based reasoning features can be employed for moral decision making, without being constrained merely to quantitative simplifying assumption (cf. MEDETHEX and ETHEL) and ILP. For instance, the role of LP abduction [27] for decision making in general is discussed in [31]. Indeed, LP abduction has been applied in a variety of areas, such as in diagnosis [24], planning [21], scheduling [29], reasoning of rational agents and decision making [32, 39], knowledge assimilation [28], natural language understanding [10], security protocols verification [1], and systems biology [42]. These applications demonstrate the potential of abduction, and it may as well be suitable for moral decision making, albeit without focusing on learning moral principles. Moral reasoning with quantitative valuation of its elements (such as actions, duties, etc.), either in utility or probability, can still be achieved with other LP-based reasoning features in combination with abduction, e.g., using preferences (see, e.g., [20]) and probabilistic LP (see, e.g., [12, 44]).
- LP provides a logic-based programming paradigm with a number of practical Prolog systems, allowing not only addressing morality issues in an abstract logical formalism (e.g., deontic logic in [40]), but also via a Prolog implementation as proof of concept and a testing ground for experimentation. The use of a theorem prover in [16] to find a proof of an operative moral principle with respect to a particular deontic logic is an attempt to provide such a testing ground for experimentation, albeit not addressing non-monotonic reasoning and moral updating concerns of Powers [40]. The use of LP, without resorting to deontic logic, to model Kant's categorical imperative and non-monotonic reasoning (via default negation), is shown in [23], but no LP updating is considered yet. To this end, a combination of LP abduction and updating may be promising in order to address moral decision making with non-monotonic reasoning and moral updating, in line with the views of Powers [40].
- While logical formalisms, such as deontic logic, permit to specify the notions of obligation, prohibition and permissibility in classic deontic operators, they are not immediately appropriate for representing morality reasoning *processes* studied in philosophy and cognitive science, such as the dual-process model of reactive

and deliberative processes [17, 22, 33, 47]. Advanced techniques in Prolog systems, such as tabling [48–50], open an opportunity to conceptually capture such processes, by appropriately applying it to considered reasoning mechanisms in moral decision making, such as LP abduction and updating.

Given this potential of LP in addressing all the above issues, there is a need to investigate further its potential. But before we do so, we need first to study more results from morality-related fields, such as philosophy and psychology, to better identify some significant moral facets which, at the start, are amenable to computational modeling by LP knowledge representation and reasoning features. This is the subject of the next chapter.

References

1. Alberti, M., Chesani, F., Gavanelli, M., Lamma, E., Torroni, P.: Security protocols verification in abductive logic programming. In: Proceedings of the 6th International Workshop on Engineering Societies in the Agents World (ESAW), LNCS, vol. 3963. Springer (2005)
2. Anderson, M., Anderson, S.L.: EthEl: Toward a principled ethical eldercare robot. In: Proceeding of the AAAI 2008 Fall Symposium on AI in Eldercare (2008)
3. Anderson, M., Anderson, S.L.: Robot be good: A call for ethical autonomous machines. Scientific American pp. 54–59 (2010)
4. Anderson, M., Anderson, S.L.: GenEth: A general ethical dilemma analyzer. In: Proceeding of the 28th AAAI Conference on Artificial Intelligence (2014)
5. Anderson, M., Anderson, S.L.: Toward ensuring ethical behavior from autonomous systems: a case-supported principle based paradigm. In: Proceeding of the AAAI Workshop on Artificial Intelligence and Ethics (1st International Workshop on AI and Ethics) (2015)
6. Anderson, M., Anderson, S.L., Armen, C.: Towards machine ethics: implementing two action-based ethical theories. In: Proceeding of the AAAI 2005 Fall Symposium on Machine Ethics (2005)
7. Anderson, M., Anderson, S.L., Armen, C.: An approach to computing ethics. IEEE Intell. Syst. **21**(4), 56–63 (2006)
8. Anderson, M., Anderson, S.L., Armen, C.: MedEthEx: a prototype medical ethics advisor. In: Proceeding of the 18th Innovative Applications of Artificial Intelligence Conference (IAAI 2006) (2006)
9. Arkoudas, K., Bringsjord, S., Bello, P.: Toward ethical robots via mechanized deontic logic. In: Proceedings of the AAAI 2005 Fall Symposium on Machine Ethics (2005)
10. Balsa, J., Dahl, V., Lopes, J.G.P.: Datalog grammars for abductive syntactic error diagnosis and repair. In: Proceedings of the Natural Language Understanding and Logic Programming Workshop (1995)
11. Baral, C.: Knowledge Representation. Reasoning and Declarative Problem Solving. Cambridge University Press, New York (2010)
12. Baral, C., Gelfond, M., Rushton, N.: Probabilistic reasoning with answer sets. Theory Pract. Logic Program. **9**(1), 57–144 (2009)
13. Beauchamp, T.L., Childress, J.F.: Principles of Biomedical Ethics. Oxford University Press, Oxford (1979)
14. Bok, S.: Lying: Moral Choice in Public and Private Life. Vintage Books, New York (1989)
15. Branting, L.K.: Reasoning with Rules and Precedents. Springer, Netherlands (2000)
16. Bringsjord, S., Arkoudas, K., Bello, P.: Toward a general logicist methodology for engineering ethically correct robots. IEEE Intell. Syst. **21**(4), 38–44 (2006)

17. Cushman, F., Young, L., Greene, J.D.: Multi-system moral psychology. In: Doris, J.M. (ed.) The Moral Psychology Handbook. Oxford University Press, Oxford (2010)
18. Dancy, J.: Moral particularism. In: E.N. Zalta (ed.) The Stanford Encyclopedia of Philosophy, Fall 2013 edn. Center for the Study of Language and Information, Stanford University http://plato.stanford.edu/archives/fall2013/entries/moral-particularism/ (2013)
19. Daniels, N.: Reflective equilibrium. In: E.N. Zalta (ed.) The Stanford Encyclopedia of Philosophy, Winter 2013 edn. Center for the Study of Language and Information, Stanford University http://plato.stanford.edu/archives/win2013/entries/reflective-equilibrium/ (2013)
20. Dell'Acqua, P., Pereira, L.M.: Preferential theory revision. J. Appl. Log. **5**(4), 586–601 (2007)
21. Eshghi, K.: Abductive planning with event calculus. In: Proceedings of the International Conference on Logic Programming. The MIT Press (1988)
22. Evans, J.S.B.T.: Thinking Twice: Two Minds in One Brain. Oxford University Press, Oxford (2010)
23. Ganascia, J.G.: Modelling ethical rules of lying with answer set programming. Eth. Inf. Technol. **9**(1), 39–47 (2007)
24. Gartner, J., Swift, T., Tien, A., Damásio, C.V., Pereira, L.M.: Psychiatric diagnosis from the viewpoint of computational logic. In: Proceedings of the 1st International Conference on Computational Logic (CL 2000), LNAI, vol. 1861, pp. 1362–1376. Springer (2000)
25. Guarini, M.: Computational neural modeling and the philosophy of ethics: reflections on the particularism-generalism debate. In: Anderson, M., Anderson, S.L. (eds.) Machine Ethics. Cambridge University Press, New York (2011)
26. Horty, J.: Agency and Deontic Logic. Oxford University Press, Oxford (2001)
27. Kakas, A., Kowalski, R., Toni, F.: The role of abduction in logic programming. In: Gabbay, D., Hogger, C., Robinson, J. (eds.) Handbook of Logic in Artificial Intelligence and Logic Programming, vol. 5. Oxford University Press, Oxford (1998)
28. Kakas, A.C., Mancarella, P.: Knowledge assimilation and abduction. In: International Workshop on Truth Maintenance, ECAI 1990 (1990)
29. Kakas, A.C., Michael, A.: An abductive-based scheduler for air-crew assignment. J. Appl. Artif. Intell. **15**(1–3), 333–360 (2001)
30. Kant, I.: Grounding for the Metaphysics of Morals, translated by Ellington, J., Hackett, Indianapolis (1981)
31. Kowalski, R.: Computational Logic and Human Thinking: How to be Artificially Intelligent. Cambridge University Press, New York, NY (2011)
32. Kowalski, R., Sadri, F.: Abductive logic programming agents with destructive databases. Ann. Math. Artif. Intell. **62**(1), 129–158 (2011)
33. Mallon, R., Nichols, S.: Rules. In: J.M. Doris (ed.) The Moral Psychology Handbook. Oxford University Press, Oxford (2010)
34. McLaren, B.M.: Extensionally defining principles and cases in ethics: an AI model. Artif. Intell. J. **150**, 145–181 (2003)
35. McLaren, B.M., Ashley, K.D.: Case-based comparative evaluation in truthteller. In: Proceedings of the 17th Annual Conference of the Cognitive Science Society (1995)
36. Muggleton, S.: Inductive logic programming. New Gener. Comput. **8**(4), 295–318 (1991)
37. Murakami, Y.: Utilitarian deontic logic. In: Proceedings of the 5th Advances in Modal Logic conference (AiML) (2004)
38. National Society of Professional Engineers (NSPE): The NSPE Ethics Reference Guide. The National Society of Professional Engineers, Alexandria, VA (1996)
39. Pereira, L.M., Dell'Acqua, P., Pinto, A.M., Lopes, G.: Inspecting and preferring abductive models. In: K. Nakamatsu, L.C. Jain (eds.) The Handbook on Reasoning-Based Intelligent Systems, pp. 243–274. World Scientific Publishers (2013)
40. Powers, T.M.: Prospects for a Kantian machine. IEEE Intell. Syst. **21**(4), 46–51 (2006)
41. Rawls, J.: A Theory of Justice. Harvard University Press, Cambridge (1971)
42. Ray, O., Antoniades, A., Kakas, A., Demetriades, I.: Abductive logic programming in the clinical management of HIV/AIDS. In: Proceeding of 17th European Conference on Artificial Intelligence. IOS Press (2006)

43. Reiter, R.: A logic for default reasoning. Artif. Intell. **13**, 81–132 (1980)
44. Riguzzi, F., Swift, T.: The PITA system: tabling and answer subsumption for reasoning under uncertainty. Theory Pract. Log. Program. **11**(4–5), 433–449 (2011)
45. Ross, W.D.: The Right and the Good. Oxford University Press, Oxford (1930)
46. Sinnott-Armstrong, W.: Consequentialism. In: E.N. Zalta (ed.) The Stanford Encyclopedia of Philosophy, Spring 2014 edn. Center for the Study of Language and Information, Stanford University (2014). http://plato.stanford.edu/archives/spr2014/entries/consequentialism/
47. Stanovich, K.E.: Rationality and the Reflective Mind. Oxford University Press, Oxford (2011)
48. Swift, T.: Tabling for non-monotonic programming. Ann. Math. Artif. Intell. **25**(3–4), 201–240 (1999)
49. Swift, T.: Incremental tabling in support of knowledge representation and resoning. Theory Pract. Log. Program. **14**(4–5), 553–567 (2014)
50. Swift, T., Warren, D.S.: XSB: extending Prolog with tabled logic programming. Theory Pract. of Log. Program. **12**(1–2), 157–187 (2012)

Chapter 3
Significant Moral Facets Amenable to Logic Programming

This chapter reports on our literature study in moral philosophy and psychology for choosing conceptual viewpoints close to LP-based reasoning. These viewpoints fall into three moral facets tackled in this book. In Sect. 3.1 we study *moral permissibility*, taking into account the Doctrines of Double Effect [20], Triple Effect [15], and Scanlonian contractualism [27]. We look into *the dual-process model* [4, 7, 17, 30], in Sect. 3.2, that stresses the interaction between deliberative and reactive processes in delivering moral decisions. Finally, in Sect. 3.3 we discuss the role of *counterfactual* thinking in moral reasoning.

3.1 Moral Permissibility

A trolley, whose conductor has fainted, is headed toward five people walking on the track. The banks of the track are so steep that these five people will not be able to get off the track in time. Hank is standing next to a switch that can turn the trolley onto a side track, thereby preventing it from killing the five people. However, there is a man standing on the side track. Hank can throw the switch, killing him; or he can refrain from doing so, letting the five die. Is it morally permissible for Hank to throw the switch?

This is a well-known moral problem, called *the trolley problem*, originally introduced in [9].[1] It concerns itself the question of moral permissibility in a dilemma involving harm. While most people tend to agree, as shown in psychological empirical tests (e.g., in [12]), that it is permissible for Hank to throw the switch for saving the five (albeit killing one), it is not trivial for people to explain the moral rules that justify their moral judgment of this permissibility.

[1] The descriptions of various trolley problem cases in this section (and Sect. 8.1) are taken from [22].

L.M. Pereira and A. Saptawijaya, *Programming Machine Ethics*,
Studies in Applied Philosophy, Epistemology and Rational Ethics 26,
DOI 10.1007/978-3-319-29354-7_3

3.1.1 The Doctrines of Double Effect and Triple Effect

It is appealing for such dilemma to explain the permissibility of Hank's judgment by resorting to utilitarianism, which favors an action that maximizes utility as being moral. It explains in this trolley problem that the action of Hank to throw the switch is permissible, as it will kill less people compared to the other consequence of letting the runaway trolley to hit the five. But then, utilitarian alone fails to explain a variant of the trolley problem (commonly known as the *Footbridge Case*), as described below, which shares the same ending:

The initial setting is similar to that of the trolley problem, in which the runaway trolley is headed toward five people walking on the track, who are unable to get off the track in time. But in this variant, Ian is on the footbridge over the trolley track, next to a heavy man, which he can shove onto the track in the path of the trolley to stop it, preventing the killing of five people. Ian can shove the man onto the track, resulting in death; or he can refrain from doing so, letting the five die. Is it morally permissible for Ian to shove the man?

For this variant, as reported also in [12], most people tend to agree that it is *not* permissible for Ian to shove the man, even though shoving the heavy man would result in the same consequence of throwing the switch in the original trolley problem (also called the *Bystander Case*), viz., saving five people albeit killing one. The issue of these moral dilemmas (and other variants of the trolley problem, discussed in Sect. 8.1) is therefore addressing the permissibility of harming one or more individuals for the purpose of saving others.

The *Doctrine of Double Effect* has been referred to explain the consistency of judgments, shared by subjects from demographically diverse populations (across gender, ethnicity, religion, age, exposure to moral coursework, etc.) on a series of moral dilemmas [11, 12, 22]. The Doctrine of Double Effect is first introduced by Thomas Aquinas in his discussion of the permissibility of self-defense [1]. The current versions of this principle all emphasize the permissibility of an action that causes a harm by distinguishing whether this harm is a mere *side-effect* of bringing about a good result, or rather an *intended means* to bringing about the same good end [20]. According to the Doctrine of Double Effect, the former action is permissible, whereas the latter is impermissible. It provides a justification to distinguish the Bystander and the Footbridge cases. According to the Doctrine of Double Effect, shoving the heavy man, which causes his death, is impermissible because it is performed as an intended means to save the five. On the other hand, throwing the switch is permissible, because the death of the man on the side track is mere a side-effect of that action.

Another principle related to the Doctrine of Double Effect is the *Doctrine of Triple Effect* [15]. This principle refines the Doctrine of Double Effect particularly on the notion about harming someone as an intended means. That is, the Doctrine of Triple Effect distinguishes further between doing an action *in order* that an effect occurs and doing it *because* that effect will occur. The latter is a new category of action, which is not accounted for in the Doctrine of Double Effect. Though the Doctrine of Triple Effect also classifies the former as impermissible, it is more tolerant to the latter (the third effect), i.e., it treats as permissible those actions performed just *because* instrumental harm will occur.

The Doctrine of Triple Effect is proposed to accommodate another case of the trolley problem, the *Loop Case*, introduced by Thomson [32], described below:

The initial setting is similar to that of the trolley problem, in which the runaway trolley is headed toward five people walking on the track, who are unable to get off the track in time. In this variant, Ned is standing next to a switch, which he can throw, that will temporarily turn the trolley onto a loop side track, which loops back towards the five. There is a heavy object on the side track. If the trolley hits the object, the object will slow the train down, giving the five people time to escape. The heavy object is a man. Ned can throw the switch, preventing the trolley from killing the five people, but killing the man. Or he can refrain from doing this, letting the five die. Is it morally permissible for Ned to throw the switch?

This case strikes most moral philosophers that throwing the switch to divert the trolley is permissible [24], though the Doctrine of Double Effect views that diverting the trolley in this Loop case as impermissible. Referring to the psychology experimental study in [12], 56 % of its respondents judged that diverting the trolley in the Loop case is permissible. To this end, Kamm [15] argues that the Doctrine of Triple Effect may provide the justification: it is permissible because it will hit the man, not in order to intentionally hit him.

But now, consider a variant of the Loop case, viz., the *Loop-Push Case* (cf. *Extra Push Case* in [15]). In the Loop-Push case the looping side track is initially empty, but it has a footbridge over it with a heavy man standing on the footbridge. While throwing the switch diverts the trolley onto the side track, an ancillary action of pushing the heavy man can be performed in order to place him on the looping side track, so as to stop the runaway trolley going back to the main track.

How can this Loop-Push case be morally distinguished from the original Loop case? In the original Loop case, the man has already been on the looping side track when the trolley is diverted, which causes the death of the man. On the other hand, in the Loop-Push case, the suffered death of the man is not merely caused by diverting the trolley, as the looping side track is initially empty. Instead, the man dies as a consequence of a further action, i.e., pushing the man. This action is intentionally performed to place the man on the side track, for the trolley to hit the man, and eventually preventing the trolley from killing the five people.

Differently from the Loop case, in this Loop-Push case the Doctrine of Triple Effect agrees with the Doctrine Double Effect that it is impermissible to save the five, as the ancillary action of pushing the heavy man onto the looping the side track serves an intended mean, in order for his heavy body to stop the trolley.

3.1.2 Scanlonian Contractualism

Recently T.M. Scanlon, a professor of moral philosophy at Harvard University, has developed a distinctive view of moral reasoning called *contractualism* [27]. It can be summarized as follows [27, p.153]:

An act is wrong if its performance under the circumstances would be disallowed by any chosen set of principles for the general regulation of behaviour that no one could reasonably reject as a basis for informed, unforced, general agreement.

The typical scope of Scanlonian contractualism is a domain of morality having to do with our duties to other people, referred to by Scanlon as "what we owe to each other", such as prohibitions against coercion, deception, harming and killing (like those exemplified by the trolley problem cases), rather than a broad moral criticism, such as premarital sex, homosexuality, etc., which are often considered immoral even they do not involve harming other people [27].

Scanlonian contractualism carries several features. First, it regards the importance of principles to provide reason for justifying an action to others, and emphasizes flexibility of moral principles. These principles may rule out some actions by ruling out the reasons on which they would be based, but they also leave wide room for interpretation, thus allowing exceptions. Indeed, as Dancy [5] points out, a formulation of a principle may be expanded and become more complex, but decisive; what looks like a conflict between principles can really be viewed as a relation between incompletely specified principles, and the matter is resolved by a more complete specification of at least one of them.

Second, reasoning is an important aspect in contractualism. That is, moral decisions are not had by merely relying on internal observations, but through a process of careful assessment that is naturally called reasoning. In Scanlonian contractualism, the method of reasoning through which we arrive at a judgment of right and wrong is a primary concern, which is further fostered in [29], so as to explain a good reason to arrive at that judgment. More specifically, a judgment of right and wrong is about the adequacy of reasons for accepting or rejecting principles under certain conditions. This is clearly expressed within Scanlonian contractualism above, which seeks some common ground that others could not reasonably reject to, thus promoting the idea of justifiability of a moral judgment. From a different angle, morality in Scanlonian contractualism can be viewed as an attempt to arrive at (possibly defeasible) argumentative consensus or agreement. Of course, sometimes people may disagree in the consensual judgment, but disagreement in judgments can be found in almost any area of inquiry, not just morality [23].

In [28], moral permissibility is addressed through the so-called *deliberative employment* of moral principles, which is in line with Scanlonian contractualism. When principles are used to guide deliberation, the question of the permissibility of actions is answered by identifying the justified but defeasible argumentative considerations, and their exceptions. This employing of moral principles is based on a view that moral dilemmas, such as the trolley problem and other similar dilemmas, typically share the same structure. They concern general principles that in some cases admit exceptions, and they raise questions about when those exceptions apply. Providing such a structure, an action is determined impermissible through deliberative employment when there is no countervailing consideration that would justify an exception to the applied general principle.

3.2 The Dual-Process Model

In the recent years, there has been a number of psychological research about the *dual-process* that apparently forms the basis of our thinking, decision making and social judgments. In psychology, the dual-process commonly refers to two types of processes, called *Type 1* and *Type 2* [7, 30], corresponding roughly to the familiar distinction between intuition and reflection. The Type 1 process is described as a fast and automatic process (it does not put a heavy load on central processing capacity), whereas Type 2 as a slow and controlled process. Further development of the dual-process theory has associated both types of processes with several attributes, e.g., experience-based decision making versus consequential decision making, associative versus rule-based, reactive versus deliberative, etc. The detailed discussion of their distinctive attributes is beyond the scope of this book, but can be referred to [7, 8].

In [30], the modes of processing in these two types are represented as a tripartite model of mind: autonomous mind for Type 1 processing, whereas Type 2 consists of the algorithmic level of processing (the algorithmic mind) plus the reflective level of processing (the reflective mind). The autonomous mind implements short-leashed goals unless overridden by the algorithmic mind, where such an overriding is initiated by higher level control, viz., higher level goal states and epistemic thinking dispositions—both exist at the reflective level of processing. Stanovich [30] also emphasizes the "cognitive decoupling" feature in Type 2 processing, which makes hypothetical thinking possible, by preventing our representations of the real world from becoming confused with representations of imaginary situations.

The dual-process model is evidenced by numerous psychological empirical tests, as discussed in [7, 30]. The role of the dual-process model in moral decision making is studied in [10], by examining the neural activity of people responding to various moral dilemmas involving physically harmful behavior, like those from the trolley problem cases. The experiments characterize each type in the dual-process with respect to applicable moral principles in these dilemmas. They found that a general principle favoring welfare-maximizing behaviors (utilitarian judgment in the Bystander case) appears to be supported by controlled cognitive processes (i.e., Type 2), whereas that prohibiting the use of harm as a means to a greater good (deontological judgment in the Footbridge case) appears to be part of the process that generates intuitive emotional responses (i.e., Type 1). When the utilitarian judgment is applied (such as in the Bystander case), the participants took longer time to make their responses. Moreover, the experiment suggests that in this case of utilitarian judgment, the Type 2 processing overrides the Type 1 coming from brain parts that produce emotion. Moral decision making is thus a product of complex interaction between these two types of the dual-process, even though each type apparently corresponds to a particular kind of moral judgment [4].

Whereas Cushman et al. [4] support the view that Type 2 is not associated with deontological (non-utilitarian) judgment, Mallon and Nichols [17] stipulate that both reasoning on moral rules and emotion (which are associated to Type 2 and 1, respectively) work together to produce non-utilitarian judgment, like in the Footbridge case. This study is based on the view that moral judgment is supported by internally represented rules and reasoning about whether particular cases fall under those rules, even in deontological (non-utilitarian) judgment, where Type 1 dominates according to [4]. It asserts that, though several studies demonstrate that people experience difficulty in justifying moral judgment generated by rapid and automatic processes (of Type 1), moral rules may still play an important role without the reasoning process being consciously accessible. Indeed, there has been no claim that Type 1 processing is non-computational, though the kind of rules that implement the Type 1 processing is not that of generally referred to in the Type 2 processing [8]. The role of moral rules, even when the reasoning process about them is inaccessible consciously, thus provides an explanation that despite this difficulty, moral judgment driven by the affective system is able to mirror (and consistent with) a particular moral rule. For example, the deontological judgment in the Footbridge case is consistent with Doctrine of Double Effect, and so is the utilitarian judgment in the Bystander case.

3.3 Counterfactual Thinking in Moral Reasoning

Counterfactual literally means contrary to the facts. Counterfactuals are conjectures about what would have happened, had an alternative event occurred. They may be stated in a variety of linguistic constructs, e.g.:

- "If only I were taller ..."
- "I could have been a winner ..."
- "I would have passed, were it not for ..."
- "Even if ..., the same consequence would have followed."

More generally, in psychology, counterfactuals are required to be in subjunctive mood rather than in indicative, which we follow in this book, as expressed by the following conditional statement[2]:

If the Antecedent *had been true, then the* Consequent *would have been true.*

Other requirements are that they must have false antecedent and they must be about particular past events [34].

Counterfactual reasoning involves thoughts on what might have been, what could have happened, had some matter—action, outcome, etc.—been different in the past.

[2]From the language construct viewpoint, this subjunctive mood is commonly known as having the *third conditionals* form [13].

It provides lessons for the future by virtue of contemplating alternatives. It permits thought debugging and supports a justification why different alternatives would have been worse or not better. It covers everyday experiences, like regret: "If only I had told her I love her!", "I should have studied harder". It may also triggers guilt responsibility, blame, and causation: "If only I had said something sooner, then I could have prevented the accident".

People typically reason about what they should or should not have done when they examine decisions in moral situation. It is therefore natural for them to engage counterfactual thoughts in such settings. As argued by Epstude and Roese [6], the function of counterfactual thinking is not just limited to the evaluation process, but occurs also in the reflection one. Through evaluation, counterfactuals help correct wrong behavior in the past, thus guiding future moral decisions. Reflection, on the other hand, permits momentary experiential simulation of possible alternatives, thereby allowing careful consideration before a moral decision is made, and to subsequently justify it.

Counterfactual theories are very suggestive of a conceptual relationship to a form of debugging, namely in view of correcting moral blame, since people ascribe abnormal antecedents an increased causal power, and are also more likely to generate counterfactuals concerning abnormal antecedents. Two distinct processes can be identified when people engage in counterfactual thinking. For one, its frequently spontaneous triggers encompass bad outcomes and "close calls" (some harm that was close to happening). Second, such thinking comprises a process of finding antecedents which, if mutated, would prevent the bad outcome from arising. When people employ counterfactual thinking, they are especially prone to change abnormal antecedents, as opposed to normal ones. Following a bad outcome, people are likely to conceive of the counterfactual "if only [some abnormal thing] had not occurred, then the outcome would not have happened". For a review on this topic, see [26].

Morality and normality judgments typically correlate. Normality mediates morality with causation and blame judgments. McCloy and Byrne [18] study the kind of counterfactual alternatives people tend to imagine, viz., those alternatives that can be controlled, in contemplating moral behaviors. The controllability in counterfactuals mediates between normality, blame and cause judgments. The importance of control, namely the possibility of counterfactual intervention, is highlighted in theories of blame that presume someone responsible only if they had some control of the outcome [33].

There has been a number of studies, both in philosophy and psychology, on the relation between causation and counterfactuals. The *counterfactual process view* of causal reasoning [19], for example, advocates counterfactual thinking as an essential part of the process involved in making causal judgments. This relation between causation and counterfactuals can be important for providing explanations in cases involving harm, which underlie people's moral cognition [31] and trigger other related questions, such as "Who is responsible?", "Who is to blame?", "Which punishment would be fair?", etc. In this book, we also explore the connection between causation and counterfactuals, focusing on agents' deliberate action, rather than on causation and counterfactuals in general. More specifically, our exploration of this topic links it to the Doctrines of Double Effect and Triple Effect and dilemmas involving harm, such as the trolley problem cases. Such cases have also been considered in psychology experimental studies concerning the role of gender and perspectives (first versus third person perspectives) in counterfactual thinking in moral reasoning, see [21]. The reader is referred to [3, 14] for a more general and broad discussion on causation and counterfactuals.

3.4 Concluding Remarks

Our study on these moral facets leads us to some considerations, indicating how these moral facets relate to one another. They set the focus of our research and will guide our exploration for identifying appropriate LP concepts to represent and reason about them.

The relevance of the Doctrines of Double Effect (DDE) and Triple Effect (DTE) is constrained in our research by their applicability to address moral permissibility. This research does not intend to cover moral facets with various deontic categories (obligation, prohibition, and permissibility) nor to formalize them using some deontic logic formalisms. Such research line has been pursued elsewhere, e.g., in [2, 25]. Moreover, our focus on this moral principles is anchored to their relevant moral examples, viz., the various cases of the trolley problem, which have been subject of experiments, whose results are readily available in the literature for validating our models.

Scanlonian contractualism concerns the *process* of moral decision making, which can sufficiently be demonstrated by an informal consensual argumentation, namely in justifying some actions or decisions in moral cases. Following Scanlonian contractualism, moral permissibility is addressed by identifying the justified but defeasible argumentative considerations. These considerations are based on moral principles, which in some cases admit exceptions. Indeed, such moral principles may be captured by the DDE and DTE to some extent. Each of these principles in itself admit some exceptions, as demonstrated by different judgments, e.g., in the Bystander and Footbridge cases with DDE as their referred moral principle. Furthermore, the third effect in DTE can be viewed as an exception to DDE, making DTE in some cases an exception to the DDE-impermissibility. It therefore seems appropriate to consider them in illustrating the argumentation process in justifying some moral decisions.

The dual-process model refers to an interaction between reactive and deliberative psychological processes in decision making. Its concept has been known close to computational procedure [16, 30]. It also has been empirically studied in psychology using, amongst others, the Bystander and Footbridge cases of the trolley problem, as reported by Greene et al. [10] and Cushman et al. [4], in Sect. 3.2. Their results, viz., concerning the different kind of processes in deontological and utilitarian judgments, will be taken as reference for finding a close LP representation to moral decision making. Another important reference for representing the interaction between the two processes is the attributes that are often associated with them, such as fast versus slow, reactive versus deliberative, etc.

Finally, besides exploring the connection between counterfactual and causation, and linking it to the Doctrine of Double and Triple Effect (as described in the previous section), we may look into its further application for justifying permissibility through argumentation processes à la Scanlonian contractualism.

All considerations in these concluding remarks are addressed in the following chapter.

References

1. Aquinas, T.: Summa Theologica II-II, Q.64, art. 7, "Of Killing". In: Baumgarth, W.P., Regan, R.J. (eds.) On Law, Morality, and Politics. Hackett, Indianapolis (1988)
2. Bringsjord, S., Arkoudas, K., Bello, P.: Toward a general logicist methodology for engineering ethically correct robots. IEEE Intell. Syst. **21**(4), 38–44 (2006)
3. Collins, J., Hall, N., Paul, L.A. (eds.): Causation and Counterfactuals. MIT Press, Cambridge, MA (2004)
4. Cushman, F., Young, L., Greene, J.D.: Multi-system moral psychology. In: Doris, J.M. (ed.) The Moral Psychology Handbook. Oxford University Press, Oxford (2010)
5. Dancy, J.: Ethics Without Principles. Oxford University Press, Oxford (2006)
6. Epstude, K., Roese, N.J.: The functional theory of counterfactual thinking. Pers. Soc. Psychol. Rev. **12**(2), 168–192 (2008)
7. Evans, J.S.B.T.: Thinking Twice: Two Minds in One Brain. Oxford University Press, Oxford (2010)
8. Evans, J.S.B.T., Stanovich, K.E.: Dual-process theories of higher cognition: advancing the debate. Perspect. Psychol. Sci. **8**(3), 223–241 (2013)
9. Foot, P.: The problem of abortion and the doctrine of double effect. Oxf. Rev. **5**, 5–15 (1967)
10. Greene, J.D., Nystrom, L.E., Engell, A.D., Darley, J.M., Cohen, J.D.: The neural bases of cognitive conflict and control in moral judgment. Neuron **44**, 389–400 (2004)
11. Hauser, M.D.: Moral Minds: How Nature Designed Our Universal Sense of Right and Wrong. Little Brown, London (2007)
12. Hauser, M., Cushman, F., Young, L., Jin, R.K., Mikhail, J.: A dissociation between moral judgments and justifications. Mind Lang. **22**(1), 1–21 (2007)
13. Hewings, M.: Advanced Grammar in Use with Answers: A Self-Study Reference and Practice Book for Advanced Learners of English. Cambridge University Press, New York (2013)
14. Hoerl, C., McCormack, T., Beck, S.R. (eds.): Understanding Counterfactuals, Understanding Causation: Issues in Philosophy and Psychology. Oxford University Press, Oxford (2011)
15. Kamm, F.M.: Intricate Ethics: Rights, Responsibilities, and Permissible Harm. Oxford University Press, Oxford (2006)
16. Kowalski, R.: Computational Logic and Human Thinking: How to be Artificially Intelligent. Cambridge University Press, New York (2011)
17. Mallon, R., Nichols, S.: Rules. In: Doris, J.M. (ed.) The Moral Psychology Handbook. Oxford University Press, Oxford (2010)
18. McCloy, R., Byrne, R.M.J.: Counterfactual thinking about controllable events. Mem. Cogn. **28**, 1071–1078 (2000)
19. McCormack, T., Frosch, C., Burns, P.: The relationship between children's causal and counterfactual judgements. In: Hoerl, C., McCormack, T., Beck, S.R. (eds.) Understanding Counterfactuals, Understanding Causation. Oxford University Press, Oxford (2011)
20. McIntyre, A.: Doctrine of double effect. In: Zalta, E.N. (ed.) The Stanford Encyclopedia of Philosophy, Fall 2011 edn. Center for the Study of Language and Information, Stanford University. http://plato.stanford.edu/archives/fall2011/entries/double-effect/ (2011)

21. Migliore, S., Curcio, G., Mancini, F., Cappa, S.F.: Counterfactual thinking in moral judgment: an experimental study. Front. Psychol. **5**, 451 (2014)
22. Mikhail, J.: Universal moral grammar: theory, evidence, and the future. Trends Cogn. Sci. **11**(4), 143–152 (2007)
23. Nagel, T.: Listening to reason. N. Y. Rev. **61**(15), 47–49 (2014)
24. Otsuka, M.: Double effect, triple effect and the trolley problem: squaring the circle in looping cases. Utilitas **20**(1), 92–110 (2008)
25. Powers, T.M.: Prospects for a Kantian machine. IEEE Intell. Syst. **21**(4), 46–51 (2006)
26. Roese, N.J.: Counterfactual thinking. Psychol. Bull. **121**(1), 133–148 (1997)
27. Scanlon, T.M.: What We Owe to Each Other. Harvard University Press, Cambridge (1998)
28. Scanlon, T.M.: Moral Dimensions: Permissibility, Meaning, Blame. Harvard University Press, Cambridge (2008)
29. Scanlon, T.M.: Being Realistic About Reasons. Oxford University Press, Oxford (2014)
30. Stanovich, K.E.: Rationality and the Reflective Mind. Oxford University Press, Oxford (2011)
31. Tetlock, P.E., Visser, P.S., Singh, R., Polifroni, M., Scott, A., Elson, S.B., Mazzocco, P., Rescober, P.: People as intuitive prosecutors: the impact of social-control goals on attributions of responsibility. J. Exp. Soc. Psychol. **43**, 195–209 (2007)
32. Thomson, J.J.: The trolley problem. Yale Law J. **279**, 1395–1415 (1985)
33. Weiner, B.: Judgments of Responsibility: A Foundation for a Theory of Social Conduct. The Guilford Press, New York (1995)
34. Woodward, J.: Psychological studies of causal and counterfactual reasoning. In: Hoerl, C., McCormack, T., Beck, S.R. (eds.) Understanding Counterfactuals, Understanding Causation. Oxford University Press, Oxford (2011)

Chapter 4
Representing Morality in Logic Programming

We start with a general Logic Programming background and notation used throughout this book, in Sect. 4.1, following the notation from [4]. The subsequent sections enumerate features in LP-based reasoning considered in this book, and discuss their appropriateness in representing various issues of moral facets elaborated in Chap. 3.

4.1 Preliminaries

By an alphabet $\mathcal{A}$ of a language $\mathcal{L}$ we mean a countable disjoint set of constants, function symbols, and predicate symbols. Moreover, an alphabet is assumed to contain a countable set of variable symbols. We use the underscore symbol (_) to specifically denote an anonymous variable. A term over $\mathcal{A}$ is defined recursively as either a variable, a constant or an expression of the form $f(t_1, \ldots, t_n)$, where f is a function symbol of $\mathcal{A}$, and t_is are terms. An *atom* over $\mathcal{A}$ is an expression of the form $p(t_1, \ldots, t_n)$, where p is a predicate symbol of $\mathcal{A}$, and t_is are terms. In this book, we write p/n to denote the predicate symbol p having arity n. A *literal* is either an atom a or its negation *not* a. Literals of the latter form is called default literals.

A term (respectively, atom and literal) is *ground* if it does not contain variables. The set of all ground terms (respectively, ground atoms) of $\mathcal{A}$ is called the Herbrand universe (respectively, Herbrand base) of $\mathcal{A}$.

Definition 4.1 (*Logic Program*) A *(normal) logic program* is a countable set of rules of the form:

$$H \leftarrow L_1, \ldots, L_m$$

where H is an atom, $m \geq 0$, and L_is ($1 \leq i \leq m$) are literals.[1]

[1]In the sequel, unless otherwise specified, we generally write *logic programs* to refer to *normal logic programs*.

L.M. Pereira and A. Saptawijaya, *Programming Machine Ethics*,
Studies in Applied Philosophy, Epistemology and Rational Ethics 26,
DOI 10.1007/978-3-319-29354-7_4

The comma operator in rules is read as conjunction. A normal logic program is called *definite* if none of its rules contains default literals. Following the standard convention, rules of the form $H \leftarrow$ are alternatively written as H. A rule of this form is called a *fact*.

The alphabet $\mathcal{A}$ used to write program P is assumed to precisely comprise all the constants, the function and predicate symbols that explicitly appear in P. By Herbrand universe (respectively, base) of P we mean the Herbrand universe (respectively, base) of $\mathcal{A}$. We denote the Herbrand base of P by $\mathcal{H}_P$. By a ground logic program we mean the set of ground rules obtained from P by substituting in all possible ways each of the variables in P by elements of its Herbrand universe.

We define next two- and three-value Herbrand interpretations and models of logic programs.[2] Let F be a set of atoms, $F = \{a_1, \ldots, a_n\}$. By *not* F we mean the set $\{not\ a_1, \ldots, not\ a_n\}$.

Definition 4.2 (*Two-valued Interpretation*) A *two-valued interpretation I* of a logic program P is a set of literals

$$I = T \cup not\ F$$

such that $T \cup F = \mathcal{H}_P$ and $T \cap F = \emptyset$.

The set T (respectively, F) is the set of atoms that are true (respectively, false) in I. The interpretation I is said two-valued because the truth value true or false is assigned to precisely one ground atom, viz., $T \cup F = \mathcal{H}_P$ and $T \cap F = \emptyset$.

Alternatively, the three-valued interpretation is defined below. It permits representing incomplete knowledge, where some atoms are neither true nor false, but rather undefined.

Definition 4.3 (*Three-valued Interpretation*) A *three-valued interpretation I* of a logic program P is a set of literals

$$I = T \cup not\ F$$

such that $T \subseteq \mathcal{H}_P$, $F \subseteq \mathcal{H}_P$ and $T \cap F = \emptyset$.

In a three-valued interpretation, the set T (respectively, F) is the set of atoms that are true (respectively, false) in I, and the truth value of the remaining atoms is undefined. Clearly, the two-valued interpretation is a special case of the three-valued one, for which $T \cup F = \mathcal{H}_P$ is additionally imposed.

We may view an interpretation I of a program P as a function $I : \mathcal{H}_P \rightarrow \mathcal{V}$, where $\mathcal{V} = \{0, 0.5, 1\}$, defined by:

$$I(A) = \begin{cases} 0 & \text{if } not\ A \in I \\ 1 & \text{if } A \in I \\ 0.5 & \text{otherwise} \end{cases}$$

[2] In the sequel, we simply write *interpretations* and *models* to refer to *Herbrand interpretations* and *Herbrand models*, respectively.

Clearly, for two-valued interpretations there is no atom A such that $I(A) = 0.5$.

Models are defined as usual, and based on a truth valuation function.

Definition 4.4 (*Truth Valuation*) If I is an interpretation, the *truth valuation* $\hat{I}$ corresponding to I is a function $\hat{I} : \mathcal{F} \rightarrow \mathcal{V}$, where $\mathcal{F}$ is the set of ground literals, conjunctions of literals, and rules formed over the language. It is defined as follows:

- If L is a ground atom, then $\hat{I}(L) = I(L)$.
- If L is a default literal, i.e., $L = not\ A$, then $\hat{I}(L) = 1 - \hat{I}(A)$.
- If S and T are conjunctions of literals, then $\hat{I}((S, T)) = min(\hat{I}(S), \hat{I}(T))$.
- If $H \leftarrow B$ is a rule, where B is a conjunction of literals, then:

$$\hat{I}(H \leftarrow B) = \begin{cases} 1 & \text{if } \hat{I}(B) \leq \hat{I}(H) \\ 0 & \text{otherwise} \end{cases}$$

For any $F \in \mathcal{F}$, the values 0, 0.5 and 1 of $\hat{I}(F)$ correspond to the truth values false, undefined and true, respectively. We write $I \models F$, for $F \in \mathcal{F}$, iff $\hat{I}(F) = 1$.

Definition 4.5 (*Model*) A interpretation I is called a (two-valued or three-valued) *model* of a program P iff for every ground instance $H \leftarrow B$ of a rule in program P we have $\hat{I}(H \leftarrow B) = 1$.

We define some orderings among interpretations and models as follows.

Definition 4.6 (*Classical Ordering* [33]) If I and J are two interpretations then we say that $I \preceq J$ if $I(A) \leq J(A)$ for any ground atom A. If $\mathcal{I}$ is a collection of interpretations, then an interpretation $I \in \mathcal{I}$ is called *minimal* in $\mathcal{I}$ if there is no interpretation $J \in \mathcal{I}$ such that $J \preceq I$ and $J \neq I$. An interpretation I is called *least* in $\mathcal{I}$ if $I \preceq J$, for any other interpretation $J \in \mathcal{I}$. A model M is called *minimal* (respectively, *least*) if it is minimal (respectively, least) among all models of P.

Definition 4.7 (*Fitting Ordering* [15]) If I and J are two interpretations then we say that $I \preceq_F J$ iff $I \subseteq J$. If $\mathcal{I}$ is a collection of interpretations, then an interpretation $I \in \mathcal{I}$ is called *F-minimal* in $\mathcal{I}$ if there is no interpretation $J \in \mathcal{I}$ such that $J \preceq_F I$ and $J \neq I$. An interpretation I is called *F-least* in $\mathcal{I}$ if $I \preceq_F J$, for any other interpretation $J \in \mathcal{I}$. A model M is called *F-minimal* (respectively, *F-least*) if it is F-minimal (respectively, F-least) among all models of P.

Note that the classical ordering is related with the *degree of truth* of their atoms, whereas the Fitting ordering is related with the *degree of information*. Under the latter ordering, the undefined value is less than both values true and false, providing that true and false being incompatible.

In [43], it is shown that every definite program has a unique least model, which determines the so-called *least model semantics* of a definite program. Other semantics for more general programs, allowing default literals in the body of a rule, have been proposed. In [17], *Stable Model Semantics* is introduced. Informally, when one assumes true some set of (hypothetical) default literals, and false all the others, some consequences follow according to the semantics of definite programs. If the consequences completely corroborate the hypotheses made, then they form a stable model. We first introduce the *Gelfond–Lifschitz* operator Γ that operates on two-valued interpretations of a program P.

Definition 4.8 (*Gelfond–Lifschitz Operator*) Let P be a logic program and I be its two-valued interpretation. The *GL-transformation of P modulo I* is the program $\frac{P}{I}$ obtained from P by performing the following operations:

- Remove from P all rules which contain a default literal $L = not\ A$ such that $\hat{I}(L) = 0$;
- Remove from all remaining rules those default literals $L = not\ A$ which satisfy $\hat{I}(L) = 1$.

Since the resulting program $\frac{P}{I}$ is a definite program, it has a unique least model J. We define $\Gamma(I) = J$.

In [17] it is shown that fixed points of the Gelfond–Lifschitz operator Γ for a program P are minimal models of P.

Definition 4.9 (*Stable Model Semantics*) A two-valued interpretation I of a logic program P is a *stable model* of P if $\Gamma(I) = I$.

Example 4.1 The program P:

$$
\begin{aligned}
a &\leftarrow not\ b.\\
b &\leftarrow not\ a.\\
c &\leftarrow not\ d.\\
d &\leftarrow not\ e.\\
p &\leftarrow a.\\
p &\leftarrow b.
\end{aligned}
$$

has two stable models:

$$I_1 = \{a, d, p, not\ b, not\ c, not\ e\} \text{ and } I_2 = \{b, d, p, not\ a, not\ c, not\ e\}.$$

One can verify that $\frac{P}{I_1}$ is the program:

$$
\begin{aligned}
a &\leftarrow .\\
d &\leftarrow .\\
p &\leftarrow a.\\
p &\leftarrow b.
\end{aligned}
$$

Therefore, $\Gamma(I_1) = I_1$.

Similarly for I_2, the program $\frac{P}{I_2}$ is:

$$\begin{aligned} b &\leftarrow . \\ d &\leftarrow . \\ p &\leftarrow a. \\ p &\leftarrow b. \end{aligned}$$

Therefore, $\Gamma(I_2) = I_2$.

Despite its advantages, that it provides semantics for more general programs than its predecessors and is closely related to autoepistemic logic and default theory (see [8, 16]), Stable Model Semantics has some drawbacks. Some programs may have no stable models, e.g., the program $p \leftarrow not\ p$. Even for programs with stable models, their semantics do not always lead to the expected intended semantics (see [4] for a discussion).

The *Well-Founded Semantics* is introduced in [44], addressing the difficulties encountered with the Stable Model Semantics. It has been shown in [34] that the Well-Founded Semantics is also equivalent to major formalizations of non-monotonic reasoning.

The Well-Founded Semantics can be viewed as *three-valued Stable Model Semantics* [35]. In order to formalize the notion of three-valued stable models, the language of programs is expanded with the additional propositional constant **u** with the property of being undefined in every interpretation. It is therefore assumed that every interpretation I satisfies:

$$\hat{I}(\mathbf{u}) = \hat{I}(not\ \mathbf{u}) = 0.5$$

A *non-negative* program is a program whose rules' bodies are either atoms or **u**. It is proven in [35] that every non-negative logic program has a unique least three-valued model.

The next definition extends the Gelfond–Lifschitz operator Γ to a three-valued operator Γ^*.

Definition 4.10 (Γ^**-operator*) Let P be a logic program and I be its three-valued interpretation. The *extended GL-transformation of P modulo I* is the program $\frac{P}{I}$ obtained from P by performing the following operations:

- Remove from P all rules which contain a default literal $L = not\ A$ such that $\hat{I}(L) = 0$;
- Replace in the remaining rules of P those default literals $L = not\ A$ which satisfy $\hat{I}(L) = 0.5$ by **u**;
- Remove from all remaining rules those default literals $L = not\ A$ which satisfy $\hat{I}(L) = 1$.

Since the resulting program $\frac{P}{I}$ is non-negative, it has a unique three-valued least model J. We define $\Gamma^*(I) = J$.

Definition 4.11 (*Well-Founded Semantics*) A three-valued interpretation I of a logic program P is a three-valued stable model of P if $\Gamma^*(I) = I$. The *Well-Founded Semantics* of P is determined by the unique F-least three-valued stable model of P, and can be obtained by the bottom-up iteration of Γ^* starting from the empty interpretation.

Example 4.2 Recall the program in Example 4.1. Let $I_0 = \emptyset$ be the empty interpretation.

- The least three-valued model of $\frac{P}{I_0}$:

$$\begin{aligned} a &\leftarrow \mathbf{u}. \\ b &\leftarrow \mathbf{u}. \\ c &\leftarrow \mathbf{u}. \\ d &\leftarrow \mathbf{u}. \\ p &\leftarrow a. \\ p &\leftarrow b. \end{aligned}$$

  is $\Gamma^*(I_0) = \{not\ e\}$.
- Let $I_1 = \Gamma^*(I_0)$. The least three-valued model of $\frac{P}{I_1}$:

$$\begin{aligned} a &\leftarrow \mathbf{u}. \\ b &\leftarrow \mathbf{u}. \\ c &\leftarrow \mathbf{u}. \\ d &\leftarrow . \\ p &\leftarrow a. \\ p &\leftarrow b. \end{aligned}$$

  is $\Gamma^*(I_1) = \{d, not\ e\}$.
- Let $I_2 = \Gamma^*(I_1)$. The least three-valued model of $\frac{P}{I_2}$:

$$\begin{aligned} a &\leftarrow \mathbf{u}. \\ b &\leftarrow \mathbf{u}. \\ d &\leftarrow . \\ p &\leftarrow a. \\ p &\leftarrow b. \end{aligned}$$

  is $\Gamma^*(I_2) = \{d, not\ c, not\ e\}$.
- Let $I_3 = \Gamma^*(I_2)$. The least three-valued model of $\frac{P}{I_3}$:

$$\begin{aligned} a &\leftarrow \mathbf{u}. \\ b &\leftarrow \mathbf{u}. \\ d &\leftarrow . \\ p &\leftarrow a. \\ p &\leftarrow b. \end{aligned}$$

  is $\Gamma^*(I_3) = \{d, not\ c, not\ e\}$.

Therefore, the well-founded model of P is $I_3 = \{d, not\ c, not\ e\}$, where d is true, c and e are both false, and a, b and p are undefined.

In the sequel, we write the well-founded model of program P as $WFM(P)$.

4.2 Abduction

The notion of abduction is first introduced by Peirce [20], and is characterized as "a step of adopting a hypothesis as being suggested by the facts".

Abduction consists of reasoning where one chooses from available hypotheses those that best explain the observed evidence, in some preferred sense. Abduction in LP is realized by extending LP with abductive hypotheses, called abducibles. An *abducible* is an atom Ab or its negation Ab^* (syntactically an atom, but denoting literal $not\ Ab$), named *positive* and *negative abducibles*, respectively, whose truth value is not initially assumed. The *negation complement of an abducible* A is denoted by $compl_{\mathcal{AB}}(A)$, where the complement of a positive abducible Ab and its negation Ab^* is defined as $compl_{\mathcal{AB}}(Ab) = Ab^*$ and $compl_{\mathcal{AB}}(Ab^*) = Ab$, respectively.

We next define an abductive framework in LP [21], which includes integrity constraints for restricting abduction. The definitions in this section are adapted from those of Alferes et al. [5].

Definition 4.12 (*Integrity Constraint*) An *integrity constraint* is a rule in the form of a denial:

$$\bot \leftarrow L_1, \ldots, L_m.$$

where $\bot/0$ is a reserved predicate symbol in $\mathcal{L}$, $m \geq 1$, and L_is $(1 \leq i \leq m)$ are literals.

Definition 4.13 (*Abductive Framework*) The triple $\langle P, \mathcal{AB}, \mathcal{IC} \rangle$ is called an *abductive framework*, where $\mathcal{AB}$ is the set of abducible predicates (and their corresponding arity), P is a logic program over $\mathcal{L} \setminus \{\bot\}$ such that there is no rule in P whose head is an abducible formed by a predicate in $\mathcal{AB}$, and $\mathcal{IC}$ is a set of integrity constraints.

Given an abductive framework $\langle P, \mathcal{AB}, \mathcal{IC} \rangle$, we write $\mathcal{AB}_{gL}$ to denote a finite set of ground abducibles formed over the set $\mathcal{AB}$.[3] In particular, $\mathcal{AB}_{gL} = \mathcal{AB}$ if the abducible predicates in $\mathcal{AB}$ are just propositional (nullary predicates).

Definition 4.14 (*Abductive Scenario*) Let F be an abductive framework $\langle P, \mathcal{AB}, \mathcal{IC} \rangle$. An *abductive scenario* of F is a tuple $\langle P, \mathcal{AB}, \mathcal{S}, \mathcal{IC} \rangle$, where $\mathcal{S} \subseteq \mathcal{AB}_{gL}$ and there is no $A \in \mathcal{S}$ such that $compl_{\mathcal{AB}}(A) \in \mathcal{S}$, i.e., $\mathcal{S}$ is consistent.

The consistency of an abductive scenario can be imposed by an integrity constraint $\bot \leftarrow Ab, Ab^*$.

[3] In practice, the body of a program rule may contain non-ground abducibles, but they have to be ground when abduced.

Let observation O be a set of literals, analogous to a query in LP. Abducing an explanation for O amounts to finding consistent abductive solutions to a goal, whilst satisfying the integrity constraints, where abductive solutions consist in the semantics obtained by replacing in P the abducibles in $\mathcal{S}$ by their truth value. We define formally abductive solutions under the Well-Founded Semantics below.

Given a scenario $\langle P, \mathcal{AB}, \mathcal{S}, \mathcal{IC} \rangle$ of an abductive framework $\langle P, \mathcal{AB}, \mathcal{IC} \rangle$, we first define $P_{\mathcal{S}}$ as the smallest set of rules that contains for each $A \in \mathcal{AB}_{gL}$, the fact A if $A \in \mathcal{S}$; and $A \leftarrow \mathbf{u}$ otherwise. Alternatively, and obviously equivalent, instead of adding to $P_{\mathcal{S}}$ the rule $A \leftarrow \mathbf{u}$, one may simply replace the corresponding A with $\mathbf{u}$ both in P and $\mathcal{IC}$.

Definition 4.15 (*Abductive Solution*) Let $F = \langle P, \mathcal{AB}, \mathcal{IC} \rangle$ and $\langle P, \mathcal{AB}, \mathcal{S}, \mathcal{IC} \rangle$ be an abductive scenario of F. The consistent set of abducibles $\mathcal{S}$ is an *abductive solution* to F if $\bot$ is false in $M_s = WFM(P \cup P_{\mathcal{S}} \cup \mathcal{IC})$. We say that $\mathcal{S}$ is an *abductive solution for query Q* if Q is true in M_s, written $M_s \models Q$.

Abduction in LP can be accomplished by a top-down query-oriented procedure for finding a query solution by need. The solution's abducibles are leaves in its procedural query-rooted call-graph, i.e., the graph is recursively generated by the procedure calls from literals in bodies of rules to heads of rules, and thence to the literals in a rule's body. The correctness of this top-down computation requires the underlying semantics to be relevant, as it avoids computing a whole model (to warrant its existence) in finding an answer to a query. Instead, it suffices to use only the rules relevant to the query—those in its procedural call-graph—to find its truth value. The Well-Founded Semantics enjoys this relevancy property, i.e., it permits finding only relevant abducibles and their truth value via the aforementioned top-down query-oriented procedure. Those abducibles not mentioned in the solution are indifferent to the query.

Representing Moral Facets by Abduction

The basic role of abduction pertains to its applicability for decision making, where abducibles are used for representing available decisions in moral dilemmas. For example, in the trolley problem, one can introduce abducibles to represent decisions like 'diverting the trolley', 'shoving the heavy man', etc., depending on the considered cases. They are abduced to satisfy a given query and integrity constraints, which reflect moral considerations in a modeled dilemma.

Other roles of abduction in representing the identified moral facets are as follows:

- With respect to moral permissibility, integrity constraints can be used for ruling out impermissible actions according to the followed moral principle. For example, when representing the Doctrine of Double Effect, integrity constraints are used for excluding those actions corresponding to intended harms for a greater good. It can therefore capture the deontological judgment with respect to the Doctrine of Double Effect. That is, integrity constraints can be applied a priori for ruling out intended harming actions, regardless how good their consequences are. On the other hand, utilitarian judgments, such as in the Bystander case, require weighing the consequences of decisions in order to arrive at a right decision that maximizing the best consequences. This is more appropriately addressed by other feature, as discussed in Sect. 4.3.

- In counterfactual reasoning, whose applications concerning moral permissibility are explored in this book, hypothetically conjecturing an alternative event requires the "other things being equal" assumption for fixing the background context of the counterfactual being evaluated. Abduction is important for providing such "other things being equal" background context. To this end, it hypothesizes incomplete information about the setting (the exogenous variables) by providing an explanation to the factual observations. This will be discussed in Chap. 6.

Related to abduction is the concept of inspection points [28], where side-effects in abduction are examined. This concept is declaratively construed with a program transformation, and procedurally constructed by 'meta-abducing' a specific abducible literal $abduced(A)$ whose function is only checking that its corresponding abducible A is indeed already abduced elsewhere. Therefore, the consequence of the action that triggers this 'meta-abducing' is merely a side-effect. In this book, the use of inspection points is only illustrated as an alternative to counterfactual reasoning in representing the Doctrine of Double Effect, by distinguishing between an instrumental cause and a mere side-effect.

For related use of inspection points to express contextual side effects and other variants of contextual abductive explanations, and their applications for modeling belief-bias effect in psychology [13, 14], the reader is referred to [29].

4.3 Preferences Over Abductive Scenarios

In abduction it is desirable to generate only abductive explanations relevant for the problem at hand. In [10], abducibles in an abductive framework are selectively assumed by introducing rules encoding domain specific information about which particular assumptions are relevant and considered in a specific situation, namely which can be instrumentally used and preferred. For this purpose, the notion of *expectation* is employed to express preconditions for an expectation of an abducible A (or its contrary), as expressed by rule $expect/1$ (or $expect\_not/1$, respectively) below:

$$\begin{aligned} expect(A) &\leftarrow L_1, \ldots, L_m. \\ expect\_not(A) &\leftarrow L_1, \ldots, L_n. \end{aligned}$$

Using this notion of expectation, an abducible A is only abduced if there is an expectation for it, and there is no expectation to the contrary. In this case, we say that the abducible A is *considered*. We discuss in Chap. 7 two possible rules defining $consider(A)$, in terms of $expect(A)$ and $expect\_not(A)$. They particularly differ in the technique employed for abducing A: one realizes abduction using an even loop over negation under Stable Model Semantics, whereas the other concerns abduction in Well-Founded Semantics (see Definition 4.15).

While the above concept of expectation constrains relevant abducibles a priori with respect to the agent's actual situation, other preferences can be enacted by examining the consequences of the considered abducibles [28]. These consequences are contained in the abductive stable models of the considered abducibles.

Definition 4.16 (*Abductive Stable Model*) Let $\langle P, \mathcal{AB}, \mathcal{IC} \rangle$ be an abductive framework and $\Delta \subseteq \mathcal{AB}_{gL}$ be a set of abducibles. An interpretation M is an *abductive stable model of* Δ if M is a stable model of $P \cup \Delta$.

Because stable models are two-valued, each abducible or its negation complement must be abduced, subject to consistency.

Since these consequences are contained in abductive stable models, such preferences can only be enacted a posteriori, viz., only after considered abducibles and their abductive stable models are computed. In evaluating consequences, a posteriori preferences can be based on quantitative measures (e.g., by utility functions) or qualitative ones (e.g., by enforcing some property over consequences to hold).

Representing Moral Facets by Preferences

In this book, a posteriori preferences are appropriate for capturing utilitarian judgment that favors welfare-maximizing behaviors. More specifically, the combined use of a priori integrity constraints and a posteriori preferences reflects an interaction of two different processes in terms of the dual-process model. Whereas a priori integrity constraints can be viewed as a mechanism to generate immediate responses in deontological judgment, reasoning with a posteriori preferences can be viewed as a form of controlled cognitive processes in utilitarian judgment. The latter is evidently a more involved process of reasoning: after excluding those abducibles that have been ruled out a priori by the integrity constraints, the consequences of the considered abducibles have first to be computed, and only then are they evaluated to prefer the solution affording the greater good.

4.4 Probabilistic LP

Moral reasoning is typically performed upon conceptual knowledge of the actions. But it is often the case that one has to pass a moral judgment on a situation without actually observing the situation, i.e., there is no full, certain information about the actions. When such uncertainty is expressed in probability values, it is relevant to probabilistically reason about actions.

A number of research has been done for integrating probability in LP, which results in a paradigm called Probabilistic LP. Many languages have been proposed in Probabilistic LP, amongst them: Independent Choice Logic [32], PRISM [39], Logic Programs with Annotated Disjunctions [45], ProbLog [36], P-log [6] and PITA [37].

These languages share similarities in their semantics, viz., using various forms of the *distribution semantics* [39]. Under the distribution semantics, a probabilistic logic program defines a probability distribution over normal logic programs (called *worlds*). The distribution is extended to a joint probability distribution over worlds and interpretations, from which the probability of a query can be obtained. The reader is referred to [38] for a survey on various Probabilistic LP under the distribution semantics.

Representing Moral Facets by Probabilistic LP

In this book, Probabilistic LP is employed as part of abduction in order to allow abducing moral decisions under uncertainty, in particular for reasoning about actions with respect to the availability of observed evidences and their attending truth value. This is relevant in moral jurisprudence, e.g., in courts, where jurors are required to proffer rulings beyond reasonable doubt based on available evidences.

The use of Probabilistic LP in court rulings illustrates a form of deliberative employment (of Scanlonian contractualism), where permissibility of actions is addressed through justified but defeasible argumentative considerations. This stance can be captured by viewing the standard probability of proof beyond reasonable doubt (for an example of such probability, see [27]) as a common ground for the verdict of guilty to be qualified as 'beyond reasonable doubt'. Argumentation may subsequently take place through presentation of other evidences with diverse level of uncertainty (these evidences are presented via LP updating, to be described in Sect. 4.5) as a consideration to justify exceptions. Whether such an evidence is accepted as a justification (by defeating the formerly presented evidence) depends on its influence on the probability of action, which in turn determines its permissibility and thereby the verdict. In other words, it depends on whether this probability is still within the agreed standard of proof beyond reasonable doubt, given that moral permissibility of actions is couched in court terms as verdicts about guilt.

4.5 LP Updating

Concomitantly to abduction an agent may learn new information from the external world or update itself internally of its own accord in order to pursue its present goal. It is therefore natural to accommodate LP abduction with updating.

LP updating allows Logic Programming to represent dynamic knowledge, i.e., knowledge that evolves over time, where updating is not only performed upon the extensional part (facts of a program), but may also take place on the intensional part (rules of a program). LP updating has been extensively studied, leading to the development of LP updating semantics [1, 3, 11, 22], a Dynamic Logic Programming framework for LP updating [3], and languages of LP updating [2, 22].

In this book, the application of LP updating is restricted to updating fluents only. Nevertheless, this restriction is sufficient for dealing with morality facets considered in this book.

Representing Moral Facets by LP Updating

LP updating permits establishing a concrete exception to a moral principle. It thus facilitates modeling Scanlon's deliberative employment for justifying permissible actions, which involves defeasible argumentation and consideration through exceptions. The role of LP updating is particularly relevant in the following moral cases:

- The knowledge of jurors in a court case, as discussed in the previous section, may be updated with new evidences, which may defeat former ones, and consequently may influence the verdict (depending on the agreed probability standard of the verdict).
- In the trolley problem, the knowledge of an agent may be updated with new information, allowing a scenario to be constructed incrementally. It thus permits demonstrating how an argumentation takes place to defend the permissibility of the agent's action in the presence of the newly obtained information.
- The agent may also adopt a new (possibly overriding) moral rule on top of those an agent currently follows, where the updating moral rule can be viewed as an exception to the current one. Such updating is necessary when the currently followed moral rule has to be revised, or qualified by an overriding exception, in the light of the situation faced by the agent.

In another role, together with abduction, LP updating forms a procedure to evaluate the validity of counterfactuals:

- Whereas abduction in the procedure provides some explanations to the factual observation, LP updating updates the causal model with a preferred explanation. It thus fixes the "other things being equal" background context of the counterfactual being evaluated into the causal model.
- LP updating also helps establish an intervention to the causal model, which is realized via a hypothetical update for representing a hypothetical alternative former event or action.

4.6 LP Counterfactuals

Counterfactual thinking in moral reasoning has been investigated particularly via psychology experiments (see, e.g., [12, 25, 26]), but it has only been limitedly explored in machine ethics. In this book, the application of counterfactual reasoning to machine ethics is fostered, while also bringing counterfactuals to a wider context of the aforementioned well-developed LP-based reasoning features by introducing a pure non-probabilistic LP-based approach for evaluating the validity of counterfactuals. It is therefore appropriate for cases when probabilities are not known or needed.

The newly introduced approach complements existing probabilistic approaches of counterfactual reasoning in LP—the latter approaches have been studied elsewhere [7, 42]—by formulating counterfactual reasoning using abduction and updating.

Representing Moral Facets by Counterfactuals

We specifically employ counterfactual reasoning to examine moral permissibility of actions according to the Doctrines of Double Effect and Triple Effect. This is achieved by distinguishing between a *cause* and a *side-effect* as a result of performing an action in order to achieve a goal. This distinction is important in explaining the permissibility of an action, both in the Doctrines of Double Effect and Triple Effect. While the issue of moral permissibility with respect to these two doctrines can be addressed using abduction with integrity constraints and a posteriori preferences (as described in Sects. 4.2 and 4.3), the use of counterfactuals provides a different approach, by means of a general counterfactual conditional form, to examine moral permissibility according to these doctrines. Furthermore, counterfactual reasoning may take place in justifying moral permissibility:

- In the form of *compound counterfactuals*, "Had I known what I know today, then if I were to have done otherwise, something preferred would have followed", for justifying with hindsight what was done in the past, in the absence of current knowledge.
- In the spirit of Scanlonian contractualism, conceptual counterfactual queries can be employed for providing a justified, but defeasible, exception to permissibility of actions.

4.7 Tabling

Tabling affords solutions reuse, rather than recomputing them, by keeping in tables subgoals and their answers obtained from query evaluation. It is now supported by a number of Prolog systems (to different extent of features), such as XSB Prolog, YAP Prolog, B-Prolog, Ciao, Mercury, and ALS Prolog. The simple idea of tabling has profound consequences [40]:

- Tabling ensures termination of programs with the *bounded term-size property*, viz., those programs where the sizes of subgoals and answers produced during an evaluation are less than some fixed number.
- Tabling can be extended to evaluate programs with negation according to the Well-Founded Semantics.
- For queries to wide classes of programs, such as datalog programs with negation, tabling can achieve the optimal complexity for query evaluation.
- Tabling integrates closely with Prolog, so that Prolog's familiar programming environment can be used, and no other language is required to build complete systems.

For a tutorial on tabling, the reader is referred to Chap. 5 of Swift et al. [41].

Representing Moral Facets by Tabling

This idea of solution reuse suggests that tabling is appropriate for an interaction of the controlled and the intuitive processes, and therefore capturing the dual-process model in moral decision-making. That is, while the controlled part of moral decision-making requires deliberative reasoning (via goal-oriented abductive reasoning), it may also rely on the tabling mechanisms, at Prolog system level, for readily obtaining solutions from tables rather than deliberatively recomputing them. The availability of such tabled solutions, managed at system level, is in line with the view of the low-level intuitive process of the dual-process model, which permits a rapid and automatic moral judgment.

In order to realize the above interaction in the dual-process model, we introduce in this book two engineering techniques for taking the benefit of tabling into two reasoning features, viz., tabling abductive solutions in contextual abduction and incremental tabling of fluents for (dual-process like) bottom-up LP updating, as well as in their integration. The resulting system is then employed to conceptually demonstrate the dual-process model in moral permissibility, where the roles of tabling in this aspect are as follows:

- As reported in [9], the dual-process model associates the intuitive process with the deontological judgment in moral dilemmas like those of the trolley problem. Given that abductive solutions represent those decisions that have been abduced according to a specific deontological moral principle (e.g., the Doctrine of Double Effect), tabling in abduction allows an agent to immediately deliver an action in a compatible context, without repeating the same deliberative reasoning. The consistency of moral decision can therefore be maintained, following that specific moral principle (as if it is obtained through deliberative reasoning), even though this decision is only directly retrieved from the table (via tabling abductive solutions).
- While the benefit of solution reuse in tabling captures the low-level and reactive parts of the dual-process model, the interaction between the deliberative and the reactive processes of this model can be demonstrated by the combination of tabling in abduction and in updating. In particular, incremental tabling may trigger an automatic (at system level) bottom-up updates propagation, where such propagation is driven by an actual abductive goal query. A scenario would be that an agent obtains some new information while making a moral decision (satisfying a given goal). In such a dynamic moral situation, where the new information is propagated by incremental tabling through its knowledge base, the agent may later be required to relaunch its previous goal or to additionally achieve some new goals in the presence of this new information and its consequences. As before, while achieving the new goals requires an extra deliberative reasoning, the decisions that have been abduced for former goals can immediately be retrieved from the table, and can subsequently be reused in the deliberative reasoning for the new goals, via contextual abduction.

4.8 Concluding Remarks

While the features discussed in this chapter are essential for our purpose of modeling the identified morality facets, it is not our strategy to prepare and develop one complex system with all features discussed. Instead, we may benefit from existing systems with some particular combined features required for modeling only some specific aspects of morality facets we discussed in this chapter.

Apart from counterfactuals and the use of tabling technique in abduction and updating, other components have initially been featured in two existing systems developed earlier, viz., ACORDA [23, 24, 31] and its subsequent development, PROBABILISTIC EPA [18, 19, 30]. These two systems include three main features: abduction, updating, and preferences. PROBABILISTIC EPA is additionally equipped with probabilistic reasoning feature based on probabilistic LP language P-log [6]. They will be discussed in Chap. 7.

We start, in the subsequent chapter, to first focus on tabling, by looking into the possibility of exploiting its mechanisms for abduction and updating, individually.

References

1. Alferes, J.J., Banti, F., Brogi, A., Leite, J.A.: The refined extension principle for semantics of dynamic logic programming. Stud. Log. **79**(1), 7–32 (2005)
2. Alferes, J.J., Brogi, A., Leite, J.A., Pereira, L.M.: Evolving logic programs. In: Proceedings of the European Conference on Artificial Intelligence (JELIA 2002). LNCS, vol. 2424, pp. 50–61. Springer, Berlin (2002)
3. Alferes, J.J., Leite, J.A., Pereira, L.M., Przymusinska, H., Przymusinski, T.: Dynamic updates of non-monotonic knowledge bases. J. Log. Program. **45**(1–3), 43–70 (2000)
4. Alferes, J.J., Pereira, L.M.: Reasoning with Logic Programming. In: LNAI, vol. 1111. Springer, Berlin (1996)
5. Alferes, J.J., Pereira, L.M., Swift, T.: Abduction in well-founded semantics and generalized stable models via tabled dual programs. Theory Pract. Log. Program. **4**(4), 383–428 (2004)
6. Baral, C., Gelfond, M., Rushton, N.: Probabilistic reasoning with answer sets. Theory Pract. Log. Program. **9**(1), 57–144 (2009)
7. Baral, C., Hunsaker, M.: Using the probabilistic logic programming language P-log for causal and counterfactual reasoning and non-naive conditioning. In: Proceedings of the 20th International Joint Conference on Artificial Intelligence (IJCAI) (2007)
8. Bidoit, N., Froidevaux, C.: General logic databases and programs: default logic semantics and stratification. J. Inf. Comput. **91**(1), 15–54 (1991)
9. Cushman, F., Young, L., Greene, J.D.: Multi-system moral psychology. In: Doris, J.M. (ed.) The Moral Psychology Handbook. Oxford University Press, Oxford (2010)
10. Dell'Acqua, P., Pereira, L.M.: Preferential theory revision. J. Appl. Log. **5**(4), 586–601 (2007)
11. Eiter, T., Fink, M., Sabbatini, G., Tompits, H.: On properties of update sequences based on causal rejection. Theory Pract. Log. Program. **6**(2), 721–777 (2002)
12. Epstude, K., Roese, N.J.: The functional theory of counterfactual thinking. Personal. Soc. Psychol. Rev. **12**(2), 168–192 (2008)
13. Evans, J.: Biases in deductive reasoning. In: Pohl, R. (ed.) Cognitive Illusions: A Handbook on Fallacies and Biases in Thinking, Judgement and Memory. Psychology Press, Hove (2012)
14. Evans, J., Barston, J.L., Pollard, P.: On the conflict between logic and belief in syllogistic reasoning. Mem. Cogn. **11**(3), 295–306 (1983)

15. Fitting, M.: A Kripke-Kleene semantics for logic programs. J. Log. Program. **2**(4), 295–312 (1985)
16. Gelfond, M.: On stratified autoepistemic theories. In: Proceedings of the 6th National Conference on Artificial Intelligence (AAAI) (1987)
17. Gelfond, M., Lifschitz, V.: The stable model semantics for logic programming. In: Proceedings of the 5th International Logic Programming Conference. MIT Press, Cambridge (1988)
18. Han, T.A.: Evolution prospection with intention recognition via computational logic. Master's thesis, Technische Universität Dresden and Universidade Nova de Lisboa (2009)
19. Han, T.A., Ramli, C.D.K., Damásio, C.V.: An implementation of extended P-log using XASP. In: Proceedings of the 24th International Conference on Logic Programming (ICLP). LNCS, vol. 5366. Springer, Heidelberg (2008)
20. Hartshorne, C., Weiss, P. (eds.): Collected Papers of Charles Sanders Peirce. Elements of Logic, vol. II. Harvard University Press, Cambridge (1932)
21. Kakas, A., Kowalski, R., Toni, F.: Abductive logic programming. J. Log. Comput. **2**(6), 719–770 (1992)
22. Leite, J.A.: Evolving Knowledge Bases: Specification and Semantics. Frontiers in Artificial Intelligence and Applications. IOS Press, Amsterdam (2003)
23. Lopes, G.: A computational approach to introspective consciousness in logic programming: ACORDA. Master's thesis, Universidade Nova de Lisboa (2006)
24. Lopes, G., Pereira, L.M.: Prospective programming with ACORDA. In: Empirically Successful Computerized Reasoning (ESCoR 2006) Workshop, IJCAR (2006)
25. McCloy, R., Byrne, R.M.J.: Counterfactual thinking about controllable events. Mem. Cogn. **28**, 1071–1078 (2000)
26. Migliore, S., Curcio, G., Mancini, F., Cappa, S.F.: Counterfactual thinking in moral judgment: an experimental study. Front. Psychol. **5**, 451 (2014)
27. Newman, J.O.: Quantifying the standard of proof beyond a reasonable doubt: a comment on three comments. Law Probab. Risk **5**(3–4), 267–269 (2006)
28. Pereira, L.M., Dell'Acqua, P., Pinto, A.M., Lopes, G.: Inspecting and preferring abductive models. In: Nakamatsu, K., Jain, L.C. (eds.) The Handbook on Reasoning-Based Intelligent Systems, pp. 243–274. World Scientific Publishers, Singapore (2013)
29. Pereira, L.M., Dietz, E.A., Hölldobler, S.: Contextual abductive reasoning with side-effects. Theory Pract. Log. Program. **14**(4–5), 633–648 (2014)
30. Pereira, L.M., Han, T.A.: Evolution prospection. In: Proceedings of the 1st KES International Symposium on Intelligent Decision Technologies (IDT), vol. 199, pp. 139–150 (2009)
31. Pereira, L.M., Lopes, G.: Prospective logic agents. Int. J. Reason. Based Intell. Syst. **1**(3/4), 200–208 (2009)
32. Poole, D.: The independent choice logic for modelling multiple agents under uncertainty. Artif. Intell. **94**(1–2), 7–56 (1997)
33. Przymusinski, T.C.: Every logic program has a natural stratification and an iterated least fixed point model. In: Proceedings of the 8th ACM Symposium on Principles Of Database Systems (PODS), pp. 11–21 (1989)
34. Przymusinski, T.C.: Three-valued non-monotonics formalisms and logic programming. In: Proceedings of the 1st International Conference on Principles of Knowledge Representation and Reasoning (KR) (1989)
35. Przymusinska, H., Przymusinski, T.C.: Semantic issues in deductive databases and logic programs. In: Formal Techniques in Artificial Intelligence: A Sourcebook, pp. 321–367. North-Holand, Amsterdam (1990)
36. Raedt, L.D., Kimmig, A., Toivonen, H.: ProbLog: a probabilistic prolog and its application in link discovery. In: Proceeidngs of the 20th International Joint Conference on Artificial Intelligence (IJCAI) (2007)
37. Riguzzi, F., Swift, T.: The PITA system: tabling and answer subsumption for reasoning under uncertainty. Theory Pract. Log. Program. **11**(4–5), 433–449 (2011)
38. Riguzzi, F., Swift, T.: Probabilistic logic programming under the distribution semantics. To appear in Festschrift in honor of David S. Warren. Available from http://coherentknowledge.com/wp-content/uploads/2013/05/plp-festschrift-paper-TS+FR.pdf (2014)

39. Sato, T.: A statistical learning method for logic programs with distribution semantics. In: Proceeidngs of the 12th International Conference on Logic Programming (ICLP) (1995)
40. Swift, T., Warren, D.S.: XSB: extending prolog with tabled logic programming. Theory Pract. Log. Program. **12**(1–2), 157–187 (2012)
41. Swift, T., Warren, D.S., Sagonas, K., Freire, J., Rao, P., Cui, B., Johnson, E., de Castro, L., Marques, R.F., Saha, D., Dawson, S., Kifer, M.: The XSB System Version 3.6.x Volume 1: Programmer's Manual (2015)
42. Vennekens, J., Bruynooghe, M., Denecker, M.: Embracing events in causal modeling: interventions and counterfactuals in CP-logic. In: JELIA. LNCS, vol. 6341, pp. 313–325. Springer, Heidelberg (2010)
43. van Emden, M.H., Kowalski, R.: The semantics of predicate logic as a programming language. J. ACM **4**(23), 733–742 (1976)
44. van Gelder, A., Ross, K.A., Schlipf, J.S.: The well-founded semantics for general logic programs. J. ACM **38**(3), 620–650 (1991)
45. Vennekens, J., Verbaeten, S., Bruynooghe, M.: Logic programs with annotated disjunctions. In: Proceedings of the 20th International Conf. on Logic Programming (ICLP). LNCS, vol. 3132. Springer, Heidelberg (2004)

Chapter 5
Tabling in Abduction and Updating

In the individual realm part of this book, we are addressing the interplay amongst appropriate LP features to represent moral facets and to reason about them. One such interplay is between LP abduction and updating, both supported with tabling mechanisms. In this chapter, we propose novel approaches for employing tabling in abduction and updating—*separately*—viz., tabling abductive solutions in contextual abduction (Sect. 5.1), plus the incremental tabling of fluents for LP updating (Sect. 5.2). Moreover, these two individual approaches form the components for their subsequent joint tabling technique when combining LP abduction and updating, to be discussed in detail in Chap. 7.

The new techniques introduced here, as well as their later joint combination, are general for normal logic programs, and not specific to morality applications. That is, they are of interest in themselves and may be applicable to other domains.

5.1 Tabling Abductive Solutions in Contextual Abduction

In abduction, it is often the case that abductive solutions found within one context are also relevant in a different context, and can be reused with little cost. As discussed in the previous chapter, the issue of reusing solutions to a goal is commonly addressed in Logic Programming—absent of abduction—by employing a tabling technique [18]. Tabling appears to be conceptually suitable for abduction too; in this case, to reuse priorly obtained abductive solutions. In practice, an abductive solution to a goal G is not immediately amenable to tabling, because such a solution is typically attached to G's abductive context. By an *abductive context of a goal G*, we mean a set of abducibles that provides the context in which an abductive solution for G is sought.[1]

[1] The set of abducibles in an abductive context is represented in the sequel using the usual Prolog list notation.

L.M. Pereira and A. Saptawijaya, *Programming Machine Ethics*,
Studies in Applied Philosophy, Epistemology and Rational Ethics 26,
DOI 10.1007/978-3-319-29354-7_5

In this section, we discuss a technique for reusing priorly obtained abductive solutions, from one abductive context to another, by benefiting from LP tabling. This technique of *tabling abductive solutions in contextual abduction*, called TABDUAL, is underpinned by ABDUAL [4], an approach for computing abduction over the Well-Founded Semantics. TABDUAL consists of a program transformation that concretely realizes the abstract theory of ABDUAL, but now in the presence of tabling abductive solutions. It specifically employs the dual program transformation, introduced in ABDUAL, to more efficiently handle the problem of abduction under negative goals.

We start by giving the motivation for the need of tabled abduction, and subsequently show how tabled abduction is conceptualized and realized in the TABDUAL transformation.

Example 5.1 (Motivation and Idea) Consider the program P_1 below in an abductive framework $\langle P_1, \{a/0, b/0\}, \emptyset \rangle$:

$$q \leftarrow a.$$
$$s \leftarrow q, b.$$
$$t \leftarrow s, q.$$

Suppose three queries are invoked, asking for individual explanations of q, s, and t, in that order.

- The first query, q, is satisfied simply by taking $[a]$ as the abductive solution for q, and tabling it.
- Executing the second query, s, amounts to satisfying the two subgoals in its body, i.e., invoking q followed by abducing b. Since q has previously been invoked, we can benefit from reusing its solution, instead of recomputing, given that the solution was tabled. That is, query s can be solved by extending the current ongoing abductive context $[b]$ of subgoal q with the already tabled abductive solution $[a]$ of q, yielding the abductive solution $[a, b]$.
- The final query t can be solved similarly. Invoking the first subgoal s results in the priorly registered abductive solution $[a, b]$, which becomes the current abductive context of the second subgoal q. Since $[a, b]$ subsumes the previously obtained (and tabled) abductive solution $[a]$ of q, we can then safely take $[a, b]$ as the abductive solution to query t.

This example shows how $[a]$, as the abductive solution of the first query q, can be reused from one abductive context of q (viz., $[b]$ in the second query, s) to another context (viz., $[a, b]$ in the third query, t). One may observe that if the body of rule q contains a huge number of subgoals, it may potentially cause an expensive recomputation of its abductive solutions, if they have not been tabled.

TABDUAL comprises two stages. The first stage is a *program transformation* that provides a self-sufficient program transform on which the second stage, the *abduction* itself, is directly enacted through queries.

5.1.1 TABDUAL *Program Transformation*

Example 5.1 indicates two key ingredients of the transformation:

- *Abductive contexts*, which relay the ongoing abductive solution from one subgoal to subsequent subgoals in the body of a rule, as well as from the head to the body of a rule, via *input* and *output* contexts.
- *Tabled predicates*, which table the abductive solutions for predicates that appear in the head of rules in the program, such that they can be reused from one abductive context to another.

The TABDUAL program transformation consists of several parts, viz., the transformations for tabling abductive solutions (Sect. 5.1.1.1), for producing dualized negation (Sect. 5.1.1.2), and for inserting abducibles into an abductive context (Sect. 5.1.1.3). This program transformation also requires a given query to be transformed. This is detailed in Sect. 5.1.1.4.

5.1.1.1 Tabling Abductive Solutions

We continue in Example 5.2 to show how to realize the idea described in Example 5.1 through a program transformation. It illustrates how every rule in P_1 is transformed, by introducing a corresponding tabled predicate with one extra argument for an abductive solution entry. The newly introduced tabled predicate therefore essentially tables this abductive solution.

Example 5.2 We show first how the rule $t \leftarrow s, q$ in P_1 is transformed into two rules:

$$t_{ab}(E_2) \leftarrow s([\,], E_1), q(E_1, E_2). \tag{5.1}$$

$$t(I, O) \leftarrow t_{ab}(E), produce\_context(O, I, E). \tag{5.2}$$

Predicate $t_{ab}(E)$ is the tabled predicate which is introduced to table one abductive solution for t in its argument E. Its definition, in the rule on the left, follows from the original definition of t. Two extra arguments, that serve as input and output contexts, are added to the subgoals s and q in the rule's body.

- Rule 5.1 expresses that the tabled abductive solution E_2 of t_{ab} is obtained by relaying the ongoing abductive solution stored in context E_1 from subgoal s to subgoal q in the body, given the empty input abductive context $[\,]$ of s. This input abductive context is empty because there is no abducible by itself in the body of the original rule of t.
- Rule 5.2 shows how the tabled abductive solution in E of t_{ab} can be reused for a given (input) abductive context of t. This rule expresses that the output abductive solution O of t is obtained from the solution entry E of t_{ab} and the given input context I of t, via the TABDUAL system predicate $produce\_context(O, I, E)$.

The system predicate $produce\_context(O, I, E)$ should guarantee that it produces a *consistent* output O from the input abductive context I and the abductive solution entry E, encompassing both.

The other two rules in P_1 are transformed following the same idea. The rule $s \leftarrow q, b$ is transformed into:

$$s_{ab}(E) \leftarrow q([b], E). \tag{5.3}$$

$$s(I, O) \leftarrow s_{ab}(E), produce\_context(O, I, E). \tag{5.4}$$

where $s_{ab}(E)$ is the predicate that tables, in E, the abductive solution of s. Notice how b, the abducible appearing in the body of the original rule of s, becomes the input abductive context of q. The same transformation is obtained, even if b comes before q in the body of the rule s.

Finally, the rule $q \leftarrow a$ is transformed into:

$$q_{ab}([a]).$$
$$q(I, O) \leftarrow q_{ab}(E), produce\_context(O, I, E).$$

where the original rule of q, which is defined solely by the abducible a, is simply transformed into the tabled fact $q_{ab}/1$.

The transformation for tabling abductive solutions is formalized in Definition 5.1.

In the sequel, we write $\bar{t}$ to denote $[t_1, \ldots, t_n]$, $n \geq 0$. For a predicate p/n, we write $p(\bar{t})$ to denote $p(t_1, \ldots, t_n)$. In particular, we write $\bar{X}$ to denote $[X_1, \ldots, X_n]$, $p(\bar{X})$ to denote $p(X_1, \ldots, X_n)$, and $p(\bar{X}, Y, Z)$ to denote $p(X_1, \ldots, X_n, Y, Z)$, where all variables are distinct.

Definition 5.1 (*Transformation: tabling abductive solutions*) Consider an abductive framework $\langle P, \mathcal{AB}, \mathcal{IC} \rangle$. By H_r and $\mathcal{B}_r$, we refer to the head and the body of rule $r \in P$, respectively. Let $\mathcal{A}_r \subseteq \mathcal{B}_r$ be the set of abducibles (either positive or negative) in $r \in P$, and r' be the rule, such that $H_{r'} = H_r$ and $\mathcal{B}_{r'} = \mathcal{B}_r \setminus \mathcal{A}_r$.

1. For every rule $r \in P$ with r' the rule $l(\bar{t}) \leftarrow L_1, \ldots, L_m$, we define $\tau'(r)$:

   $$l_{ab}(\bar{t}, E_m) \leftarrow \alpha(L_1), \ldots, \alpha(L_m).$$

   where α is defined as:

   $$\alpha(L_i) = \begin{cases} l_i(\bar{t}_i, E_{i-1}, E_i) & \text{, if } L_i \text{ is an atom } L_i = l_i(\bar{t}_i) \\ not\_l_i(\bar{t}_i, E_{i-1}, E_i) & \text{, if } L_i \text{ is a default literal } L_i = not\ l_i(\bar{t}_i) \end{cases}$$

   with $1 \leq i \leq m$, E_i are fresh variables and $E_0 = \mathcal{A}_r$. Note that variables E_is serve as *abductive contexts*.
2. For every predicate p/n with rules in P (i.e., rules whose heads is the predicate p/n), we define $\tau^+(p)$:

$$p(\bar{X}, I, O) \leftarrow p_{ab}(\bar{X}, E),\ produce\_context(O, I, E).$$

where $produce\_context(O, I, E)$ is a TABDUAL system predicate that concerns itself with: whether E is already contained in I and, if not, whether there are any abducibles from E, consistent with I, that can be added to produce O. If E is inconsistent with I then the specific entry E cannot be reused with I, $produce\_context/3$ fails and another entry E is sought.

Example 5.3 Consider an abductive framework $\langle P, \{a/1\}, \emptyset\rangle$, where the program P (whose rules are named with r_i) is given below:

$$r_1 : u(0, \_).$$
$$r_2 : u(s(X), Y) \leftarrow a(X), v(X, Y, Z), not\ w(Z).$$
$$r_3 : v(X, X, s(X)).$$

We have $\mathcal{A}_{r_i}$ and r'_i, for $1 \leq i \leq 3$, as follows:

- $\mathcal{A}_{r_1} = [\,]$ and $r'_1 :\ u(0, \_)$.
- $\mathcal{A}_{r_2} = [a(X)]$ and $r'_2 :\ u(s(X), Y) \leftarrow v(X, Y, Z), not\ w(Z)$.
- $\mathcal{A}_{r_3} = [\,]$ and $r'_3 :\ v(X, X, s(X))$.

The transformation of Definition 5.1 results in:

$$\tau'(r_1) : u_{ab}(0, \_, [\,]).$$
$$\tau'(r_2) : u_{ab}(s(X), Y, E_2) \quad \leftarrow v(X, Y, Z, [a(X)], E_1),\ not\_w(Z, E_1, E_2).$$
$$\tau'(r_3) : v_{ab}(X, X, s(X), [\,]).$$

$$\tau^+(u) : u(X_1, X_2, I, O) \quad \leftarrow u_{ab}(X_1, X_2, E),\ produce\_context(O, I, E).$$
$$\tau^+(v) : v(X_1, X_2, X_3, I, O) \ \leftarrow v_{ab}(X_1, X_2, X_3, E),\ produce\_context(O, I, E).$$

Observe that both arguments of $u/2$ are kept in the tabled predicate u_{ab} (as its first two arguments), and one extra argument is added (as its third argument) for tabling its abductive solution entry. Similar transformation also applies to $v/3$. The transformation does not create (and indeed do not need) $\tau^+(w)$, because there is no rule whose head is $w/1$ in the program P.

5.1.1.2 Abduction Under Negative Goals

For abduction under negative goals, the program transformation employs the *dual program transformation* of ABDUAL [4]. In this transformation, negative goals are syntactically treated as new atoms. The motivation behind the transformation is to enable us to obtain the solutions of a negative goal *not* G without having to compute all abductive solutions of the positive goal G and subsequently to negate their disjunction. In other words, with the dual program transformation, the abductive solutions of a negative goal can be obtained one at a time, as we treat abduction under positive goals.

The idea of the dual program transformation is to define, for each atom A and its (possibly empty) set of rules $\mathcal{R}_A$ in a program P, a set of dual rules whose head is not_A, such that not_A is true if and only if A is false by $\mathcal{R}_A$ in the employed semantics of P. Note that, instead of having a negative goal *not* A as the head of a dual rule, we use its corresponding atom not_A. Example 5.4 illustrates this main idea of the dual transformation is realized in TABDUAL by means of two-layer dual rules.

Example 5.4 Consider an abductive framework $\langle P_2, \{a/0\}, \emptyset\rangle$, where the program P_2 is as follows:

$$p \leftarrow a. \tag{5.5}$$

$$p \leftarrow q, not\ r. \tag{5.6}$$

$$r.$$

- With regard to atom p, the transformation creates a set of dual rules for p which falsify p with respect to its two rules. That is, both rules 5.5 *and* 5.6 are falsified, as expressed below by predicate p^{*1} and p^{*2}, respectively:

  $$not\_p(T_0, T_2) \leftarrow p^{*1}(T_0, T_1),\ p^{*2}(T_1, T_2).$$

  We refer this resulting single rule as the *first layer* of the dual program transformation (or, the first layer dual rule). In this rule, input and output abductive context arguments, T_0 and T_2, respectively, are added in the head. Similarly, these context arguments are added into each subgoal of the rule's body, where intermediate context T_1 relays the ongoing abductive solution from p^{*1} to p^{*2}.
  The *second layer* contains the definitions of p^{*1} and p^{*2}, where p^{*1} and p^{*2} are defined by falsifying the body of p's first rule (5.5) and second one (5.6), respectively.

  - In case of p^{*1}, rule 5.5 is falsified only by abducing the negation of the abducible a. Therefore, we have:

    $$p^{*1}(I, O) \leftarrow a^*(I, O).$$

    Notice that the negation a^* of the abducible a refers to its own abduction, which is achieved by invoking the subgoal $a^*(I, O)$. This subgoal is defined via the transformation of abducibles, which will be discussed in Sect. 5.1.1.3.
  - In case of p^{*2}, rule 5.6 is falsified by alternatively failing one subgoal in its body at a time, viz., by negating q or by negating *not* r:

    $$p^{*2}(I, O) \leftarrow not\_q(I, O). \qquad p^{*2}(I, O) \leftarrow r(I, O).$$

- With regard to atom q, the dual program transformation produces the fact:

  $$not\_q(I, I).$$

as its dual rule for an obvious reason, that there is no rule with the head q in P_2. As a fact, the content of the context I in this dual rule is simply relayed from the input to the output abductive context. That is, having an empty body, the output abductive context does not depend on the context of any other goals, but depends only on its corresponding input abductive context.

- With regard to atom r, since r is a fact, in principle it produces the first layer rule:

$$not\_r(T_0, T_1) \leftarrow r^{*1}(T_0, T_1).$$

but with no definition of $r^{*1}/2$. In other words, invoking $not\_r(\_, \_)$ will vacuously fail.

This example particularly shows how the dual rules for nullary predicates are derived, viz., by falsifying the bodies of their corresponding positive rules. In the case of non-nullary predicates, a goal may also fail (or equivalently, its negation succeeds), when its arguments disagree with the arguments of its rules. For instance, if we have just a fact $q(1)$, then goal $q(0)$ will fail (or equivalently, goal *not* $q(0)$ succeeds). It therefore provides some guidance on how to treat non-nullary predicates in the dual program transformation. That is, besides falsifying the body of a rule, a dual of a non-nullary predicate can additionally be defined by disunifying its arguments and the arguments of its corresponding source rule, as illustrated in Example 5.5.

Example 5.5 Consider an abductive framework $\langle P_3, \{a/1\}, \emptyset \rangle$, where the program P_3 is as follows:

$$q(0). \tag{5.7}$$

$$q(s(X)) \leftarrow a(X). \tag{5.8}$$

The dual program transformation of non-nullary predicate $q/1$ is given below:

1. $not\_q(X, T_0, T_2) \leftarrow q^{*1}(X, T_0, T_1), q^{*2}(X, T_1, T_2).$
2. $q^{*1}(X, I, I) \leftarrow X \backslash = 0.$
3. $q^{*2}(X, I, I) \leftarrow X \backslash = s(\_).$
4. $q^{*2}(s(X), I, O) \leftarrow a^{*}(X, I, O).$

Line 1 shows the first layer dual rule for predicate $q/1$, which is defined as usual, i.e., $q/1$ is falsified by falsifying both rules 5.7 and 5.8. Lines 2–4 show the second layer dual rules for $q/1$:

- In case of q^{*1}, the first rule of $q/1$, which is fact $q(0)$, is falsified by disunifying q^{*1}'s argument X with 0 (line 2). This is the only way to falsify rule 5.7, since it has no body.
- In case of q^{*2}, the second rule of $q/1$ is falsified by disunifying q^{*2}'s argument X with the term $s(\_)$ (line 3). Or, it can alternatively be falsified by keeping the head's argument, but falsifying its body, i.e., by abducing the negation of the abducible $a/1$ (line 4).

We next specify, in Definition 5.2, the transformation that constructs the two-layer dual rules in TABDUAL.

Definition 5.2 (*Transformation: constructing dual rules*) Consider an abductive framework $\langle P, \mathcal{AB}, \mathcal{IC}\rangle$. Let P^+ be $P \cup \mathcal{IC}$.

1. For every predicate p/n, $n \geq 0$, whose rules in P^+ are as follows:

$$p(\bar{t}_1) \leftarrow L_{11}, \ldots, L_{1n_1}.$$
$$\vdots$$
$$p(\bar{t}_m) \leftarrow L_{m1}, \ldots, L_{mn_m}.$$

with $n_i \geq 0$, $1 \leq i \leq m$:

   a. The first layer dual rule is defined by $\tau^-(p)$:

$$not\_p(\bar{X}, T_0, T_m) \leftarrow p^{*1}(\bar{X}, T_0, T_1), \ldots, p^{*m}(\bar{X}, T_{m-1}, T_m).$$

   with T_i, $0 \leq i \leq m$, are fresh variables. Note that variables T_is serve as abductive contexts.

   b. The second layer dual rules are defined by:
   $\tau^*(p) = \bigcup_{i=1}^{m} \tau^{*i}(p)$ and $\tau^{*i}(p)$ is the smallest set that contains the following rules:

$$p^{*i}(\bar{X}, I, I) \leftarrow \bar{X} \neq \bar{t}_i.$$
$$p^{*i}(\bar{t}_i, I, O) \leftarrow \sigma(L_{i1}, I, O).$$
$$\vdots$$
$$p^{*i}(\bar{t}_i, I, O) \leftarrow \sigma(L_{in_i}, I, O).$$

   where σ is defined as follows:

$$\sigma(L_{ij}, I, O) = \begin{cases} l_{ij}(\bar{t}_{ij}, I, O) & \text{, if } L_i \text{ is a default literal } not\ l_{ij}(\bar{t}_{ij}) \text{ or} \\ & \quad \text{a negative abducible } l_{ij}{}^*(\bar{t}_{ij}) \\ not\_l_{ij}(\bar{t}_{ij}, I, O) & \text{, if } L_i \text{ is an atom } l_{ij}(\bar{t}_{ij}) \\ l_{ij}{}^*(\bar{t}_{ij}, I, O) & \text{, if } L_i \text{ is a positive abducible } l_{ij}(\bar{t}_{ij}) \end{cases}$$

   Notice that, in case of $p/0$ (i.e. $n = 0$), rule $p^{*i}(\bar{X}, I, I) \leftarrow \bar{X} \neq \bar{t}_i$ is omitted, since both $\bar{X}$ and $\bar{t}_i$ are [ ]. This means, when $p/0$ is defined as a fact in P^+, we have $not\_p(T_0, T_1) \leftarrow p^{*1}(T_0, T_1)$ in the first layer, but there is no rule of $p^{*1}/2$ in the second layer (cf. the dual rule of predicate $r/0$ in Example 5.4).

2. For every predicate r/n in P^+ ($n \geq 0$) that has no rule, we define $\tau^-(r)$:

$$not\_r(\bar{X}, I, I).$$

   In particular, if $\mathcal{IC} = \emptyset$, we have $\tau^-(\bot) : not\_\bot(I, I)$.

Example 5.6 Recall Example 5.3. The transformation of Definition 5.2 results in:

$$\begin{aligned}
\tau^-(u) &: not_u(X_1, X_2, T_0, T_2) \leftarrow u^{*1}(X_1, X_2, T_0, T_1),\ u^{*2}(X_1, X_2, T_1, T_2).\\
\tau^-(v) &: not_v(X_1, X_2, X_3, T_0, T_1) \leftarrow v^{*1}(X_1, X_2, X_3, T_0, T_1).\\
\tau^-(w) &: not_w(X, I, I).\\
\tau^-(\bot) &: not_\bot(X, I, I).
\end{aligned}$$

$$\begin{aligned}
\tau^*(u) &: u^{*1}(X_1, X_2, I, I) \leftarrow [X_1, X_2] \neq [0, _].\\
&\quad u^{*2}(X_1, X_2, I, I) \leftarrow [X_1, X_2] \neq [s(X), Y].\\
&\quad u^{*2}(s(X), Y, I, O) \leftarrow a^*(X, I, O).\\
&\quad u^{*2}(s(X), Y, I, O) \leftarrow not_v(X, Y, Z, I, O).\\
&\quad u^{*2}(s(X), Y, I, O) \leftarrow w(Z, I, O).\\
\tau^*(v) &: v^{*1}(X_1, X_2, X_3, I, I) \leftarrow [X_1, X_2, X_3] \neq [X, X, s(X)].
\end{aligned}$$

5.1.1.3 Transforming Abducibles

In Example 5.4, $p^{*1}(I, O)$ is defined by abducing a^*, achieved by invoking subgoal $a^*(I, O)$. Abduction in TABDUAL is realized by transforming each abducible into a rule that updates the abductive context with the transformed abducible. For instance, abducible a of Example 5.4 translates to:

$$a(I, O) \leftarrow insert\_abducible(a, I, O).$$

where $insert\_abducible/3$ is a TABDUAL system predicate that inserts the given abducible into the abductive context, described in Definition 5.3 below. Abducible a^* is transformed similarly.

The specification for the transformation of abducibles is given in Definition 5.3.

Definition 5.3 (*Transformation of abducibles*) Given an abductive framework $\langle P, \mathcal{AB}, \mathcal{IC}\rangle$. For every $a/n \in \mathcal{AB}$, we define $\tau^\circ(a)$ as the smallest set that contains the rules:

$$\begin{aligned}
a(X_1, \ldots, X_n, I, O) &\leftarrow insert_abducible(a(X_1, \ldots, X_n), I, O).\\
a^*(X_1, \ldots, X_n, I, O) &\leftarrow insert_abducible(a^*(X_1, \ldots, X_n), I, O).
\end{aligned}$$

where $insert\_abducible(A, I, O)$ is a TABDUAL system predicate a TABDUAL system that inserts the abducible A into the input context I, resulting in the output context O. It maintains the consistency of the abductive context, failing if inserting A results in an inconsistent one.

Example 5.7 Recall Example 5.3. The transformation of Definition 5.3 results in:

$$\begin{aligned}
\tau^\circ(a(X)) : a(X, I, O) &\leftarrow insert_abducible(a(X), I, O).\\
a^*(X, I, O) &\leftarrow insert_abducible(a^*(X), I, O).
\end{aligned}$$

The specification of the complete TABDUAL program transformation is given in Definition 5.4.

Definition 5.4 (TABDUAL *transformation*) Let $\mathcal{F}=\langle P, \mathcal{AB}, \mathcal{IC}\rangle$ be an abductive framework, $\mathcal{P}$ be the set of predicates in P and $P^+ = P \cup \mathcal{IC}$. Taking:

- $\tau'(\mathcal{F}) = \{\tau'(r) \mid r \in P\}$
- $\tau^+(\mathcal{F}) = \{\tau^+(p) \mid p \in \mathcal{P}$ with rules in $P\}$
- $\tau^-(\mathcal{F}) = \{\tau^-(p) \mid p \in \mathcal{P} \cup \{\bot\}\}$
- $\tau^*(\mathcal{F}) = \{\tau^*(p) \mid p \in \mathcal{P} \cup \{\bot\}$ with rules in $P^+\}$
- $\tau^\circ(\mathcal{F}) = \{\tau^\circ(a) \mid a \in \mathcal{AB}\}$

The TABDUAL transformation $\tau(\mathcal{F})$ is defined as:

$$\tau(\mathcal{F}) = \tau'(\mathcal{F}) \cup \tau^+(\mathcal{F}) \cup \tau^-(\mathcal{F}) \cup \tau^*(\mathcal{F}) \cup \tau^\circ(\mathcal{F})$$

Example 5.8 The set of rules obtained in Examples 5.3, 5.6, and 5.7 forms $\tau(\mathcal{F})$ of the abductive framework $\mathcal{F}=\langle P, \mathcal{AB}, \mathcal{IC}\rangle$.

5.1.1.4 Transforming Queries

A query to a program, consequently, should be transformed:

- A positive goal G is simply augmented with the two extra arguments for the input and output abductive contexts.
- A negative goal *not G* is renamed into *not_G*, and added the two extra (input and output) abductive context arguments.

Moreover, a query should additionally ensure that all integrity constraints are satisfied. Note that when there is no integrity constraint, then, following Definition 5.2, the following fact is added into the transform:

$$not\_\bot(I, I).$$

Otherwise, integrity constraints are transformed just like any other rules, omitting the transformed rules with the heads $\bot_{ab}(E)$ and $\bot(I, O)$.

Finally, a query should always be conjoined with $not\_\bot/2$ to ensure that all integrity constraints are satisfied.

Example 5.9 Query

$$?\text{-}\ \ not\ p.$$

first transforms into $not\_p(I, O)$. Then, to satisfy all integrity constraints, it is conjoined with $not\_\bot/2$, resulting in top goal:

$$?\text{-}\ \ not\_p([\ ], T),\ \ not\_\bot(T, O).$$

where O is an abductive solution to the query, given initially an empty input abductive context. Note, how O is obtained by further constraining the output abductive context T for *not_p*, via passing it to the subsequent subgoal $not\_\bot$ for confirmation.

Definition 5.5 provides the specification of the query transformation.

Definition 5.5 (*Transformation of queries*) Let $\langle P, \mathcal{AB}, \mathcal{IC}\rangle$ be an abductive framework and Q be a query:

$$?\text{-}\ \ G_1, \ldots, G_m.$$

TABDUAL transforms query Q into $\Delta(Q)$:

$$?\text{-}\ \ \delta(G_1), \ldots, \delta(G_m), not\_\bot(T_m, O).$$

where δ is defined as:

$$\delta(G_i) = \begin{cases} g_i(\bar{t}_i, T_{i-1}, T_i) & \text{, if } G_i = g_i(\bar{t}_i) \\ not\_g_i(\bar{t}_i, T_{i-1}, T_i) & \text{, if } G_i = not\ g_i(\bar{t}_i) \end{cases}$$

T_0 is a given initial abductive context (or an empty context [ ], by default), $1 \le i \le m$, T_i, O are fresh variables. The output abductive context O returns the abductive solution(s) of the query.

Example 5.10 Recall Example 5.3. Query:

$$?\text{-}\ \ u(0, s(0)),\ not\ u(s(0), 0).$$

is transformed by Definition 5.5 into:

$$?\text{-}\ \ u(0, s(0), [\ ], T_1),\ not\_u(s(0), 0, T_1, T_2),\ not\_\bot(T_2, O).$$

5.1.2 Implementation Aspects

TABDUAL is implemented in XSB Prolog [21], and many of its implementation aspects benefit from features of XSB Prolog.

5.1.2.1 Grounding Dualized Negated Subgoals

Example 5.11 Consider an abductive framework $\langle P_4, \{a/1\}, \mathcal{IC}_4\rangle$, where the program P_4 is as follows:

$$q(1).$$
$$r(X) \leftarrow a(X).$$

and $\mathcal{IC}_4$:

$$\bot \leftarrow q(X), r(X).$$

The TABDUAL transformation results in:

$$\begin{array}{lll}
1. & q_{ab}(1, [\,]). & \\
2. & q(X, I, O) & \leftarrow q_{ab}(X, E), produce_context(O, I, E). \\
3. & not_q(X, I, O) & \leftarrow q^{*1}(X, I, O). \\
4. & q^{*1}(X, I, I) & \leftarrow X \backslash = 1. \\
 & & \\
5. & r_{ab}(X, [a(X)]). & \\
6. & r(X, I, O) & \leftarrow r_{ab}(X, E), produce_context(O, I, E). \\
7. & not_r(X, I, O) & \leftarrow r^{*1}(X, I, O). \\
8. & r^{*1}(X, I, I) & \leftarrow X \backslash = _. \\
9. & r^{*1}(X, I, O) & \leftarrow a^{*}(X, I, O). \\
 & & \\
10. & not_\bot(I, O) & \leftarrow \bot^{*1}(I, O). \\
11. & \bot^{*1}(I, O) & \leftarrow not_q(X, I, O). \\
12. & \bot^{*1}(I, O) & \leftarrow not_r(X, I, O).
\end{array}$$

Consider query $q(1)$, which is transformed into:

$$?\text{-}\quad q(1, [\,], T), not\_\bot(T, O).$$

Invoking the first subgoal, $q(1, [\,], T)$, results in $T = [\,]$. Invoking subsequently the second subgoal, $not\_\bot([\,], O)$, results in the abductive solution of the given query: $O = [a^{*}(X)]$, obtained via rules 10, 12, 7, and 9. Note that rule 11, an alternative to $\bot^{*1}$, fails due to uninstantiated X in its subgoal $not\_q(X, I, O)$, which leads to failing rules 3 and 4. For the same reason, rule 8, an alternative to r^{*1}, also fails.

Instead of having $[a^{*}(1)]$ as the abductive solution to the query $q(1)$, we have a non-ground abductive solution $[a^{*}(X)]$. It does not meet our requirement, in Sect. 4.2, that abducibles must be ground on the occasion of their abduction. The problem can be remedied by instantiating X, in rule 12, thereby eventually grounding the abducible $a^{*}(X)$ when it is abduced, i.e., the argument X of subgoal $a^{*}/3$, in rule 9, becomes instantiated.

In the implementation, grounding a dualized negated subgoal is achieved as follows: in addition to placing a negated literal, say *not_p*, in the body of the second layer dual rule, all positive literals that precede literal p, in the body of the corresponding source rule, are also kept in the body of the dual rule. For rule 12, introducing the positive subgoal $q(X)$, originating from the source rule, before the negated subgoal $not\_r(X, I, O)$ in the body of rule 12, helps instantiate X in this case. Rule 12 now becomes (all other rules remain the same):

$$\bot^{*1}(I, O) \leftarrow q(X, I, T), not\_r(X, T, O).$$

Notice that, differently from before, the rule is now defined by introducing all positive literals that appear before r in the original rule; in this case we introduce $q/3$ before $not\_r/3$. As the result, the argument X in $not\_r/3$ is instantiated to 1, due to the invocation of $q/3$, just like the case in the source rule. It eventually helps ground the negated abducible $a^*(X)$, when it is abduced, and the correct abductive solution $[a^*(1)]$ to query $q(1)$ is returned. By implementing this technique, we are also able to deal with non-ground positive goals, e.g., query $q(X)$ gives the correct abductive solution as well, i.e. $[a^*(1)]$ for $X = 1$.

There are some points to remark on regarding this implementation technique:

- The semantics of dual rules does not change because the conditions for failure of their source counterpart rules are that one literal must fail, even if the others succeed. The cases where the others do not succeed are handled in the other alternatives of dual rules.
- This technique may benefit from the TABDUAL's tabled predicate, e.g. q_{ab} for predicate q, as it helps avoid redundant derivations of the newly introduced positive literals in dual rules.
- Information about shared variables in the body and whether they are local or not, may be useful to avoid introducing positive literals that are not contributing to further grounding.

5.1.2.2 Dealing with Non-Ground Negative Goals

Example 5.12 Consider an abductive framework $\langle P_5, \{a/1\}, \emptyset\rangle$, where the program P_5 is as follows:

$$p(1) \leftarrow a(1).$$
$$p(2) \leftarrow a(2).$$

Query $p(X)$ succeeds under TABDUAL, giving two abductive solutions, viz., $[a(1)]$ and $[a(2)]$ for $X = 1$ and $X = 2$, respectively. But query *not* $p(X)$ does not deliver the expected solution. Instead of returning the abductive solution $[a^*(1), a^*(2)]$ for any instantiation of X, it returns $[a^*(1)]$ for a particular $X = 1$. In order to find the problem, we first look into the definition of $not\_p/3$:

$$\begin{array}{lll}
1. & not_p(X, I, O) & \leftarrow p^{*1}(X, I, T), p^{*2}(X, T, O).\\
2. & p^{*1}(X, I, I) & \leftarrow X \backslash= 1.\\
3. & p^{*1}(1, I, O) & \leftarrow a^*(1, I, O).\\
4. & p^{*2}(X, I, I) & \leftarrow X \backslash= 2.\\
5. & p^{*2}(2, I, O) & \leftarrow a^*(2, I, O).
\end{array}$$

Recall that query ?- *not* $p(X)$ is transformed into:

$$?\text{-}\ not\_p(X, [\,], N), not\_\bot(N, O).$$

When the goal $not\_p(X, [\,], N)$ is launched, it first invokes $p^{*1}(X, [\,], T)$. It succeeds by the second rule of p^{*1}, in line 3 (the first rule, in line 2, fails it), with variable X is instantiated to 1 and T to $[a^*(1)]$. The second subgoal of $not\_p(X, [\,], N)$ is subsequently invoked with the same instantiation of X and T, i.e. $p^{*2}(1, [a^*(1)], O)$, and it succeeds by the first rule of p^{*2}, in line 4, and results in $N = [a^*(1)]$. Since there is no integrity constraint, i.e., $\mathcal{IC}_5 = \emptyset$, the abductive solution $[a^*(1)]$ is just relayed from N to O, due to the dual $not\_\bot(I, I)$ in the transformed program (see Definition 5.2), thus returning the abductive solution $[a^*(1)]$ for the original query ?- *not* $p(X)$, where X is instantiated to 1.

The culprit is that both subgoals of $not\_p/3$, viz., $p^{*1}/3$ and $p^{*2}/3$, share the argument X of $p/1$. This should not be the case, as $p^{*1}/3$ and $p^{*2}/3$ are derived from two different rules of $p/1$, hence failing p should be achieved by invoking p^{*1} and p^{*2} with an independent argument X. In other words, different variants of the calling argument X should be used in $p^{*1}/3$ and $p^{*2}/3$, as shown for rule $not\_p/3$ (line 1) below:

$$\begin{aligned} not\_p(X, T_0, T_2) \leftarrow\ & copy\_term([X], [X_1]), p^{*1}(X_1, T_0, T_1), \\ & copy\_term([X], [X_2]), p^{*2}(X_2, T_1, T_2). \end{aligned}$$

where the Prolog built-in predicate $copy\_term/2$ provides a variant of the list of arguments; in this example, we simply have only one argument, i.e. $[X]$.

Now, $p^{*1}/3$ and $p^{*2}/3$ are invoked using variant independent calling arguments, viz., X_1 and X_2, respectively. The same query first invokes $p^{*1}(X_1, [\,], T_1)$, which results in $X_1 = 1$ and $T_1 = [a^*(1)]$ (by rule 3), and subsequently invokes $p^{*2}(X_2, [a^*(1)], T_2)$, resulting in $X_2 = 2$ and $T_2 = [a^*(1), a^*(2)]$ (by rule 5). It eventually ends up with the expected abductive solution: $[not\ a(1), not\ a(2)]$ for any instantiation of X, i.e., X remains unbound.

The technique ensures, as this example shows, that query ?- $p(X)$ fails for every X, and its negation, ?- *not* $p(X)$, hence succeeds. The dual rules produced for the negation are tailored to be, by definition, an 'if and only if' with regard to their corresponding source rules. If we added the fact $p(Y)$ to P_5, then the same query ?- *not* $p(X)$ would not succeed because now we have the first layer dual rule:

$$\begin{aligned} not\_p(X, T_0, T_3) \leftarrow\ & copy\_term([X], [X_1]), p^{*1}(X_1, T_0, T_1), \\ & copy\_term([X], [X_2]), p^{*2}(X_2, T_1, T_2), \\ & copy\_term([X], [X_3]), p^{*3}(X_3, T_2, T_3). \end{aligned}$$

and an additional second layer dual rule $p^{*3}(X, \_, \_) \leftarrow X \neq \_$ that always fails; its abductive contexts are thus irrelevant.

5.1.2.3 Transforming Predicates Comprising Just Facts

TABDUAL transforms predicates that comprise just facts as any other rules in the program. For instance, see fact $q(1)$ and its transformed rules (rules 1–4), in

Example 5.11. This is clearly superfluous as facts do not induce any abduction. For programs with large factual data, a simpler transformation for predicates that consist of just facts is desirable.

Suppose a predicate $q/1$ consists of just facts:

$$q(1). \quad q(2). \quad q(3).$$

Rather than applying the transformation defined in Definition 5.1, rules $q_{ab}/2$ and $q/3$ can be substituted by a single rule:

$$q(X, I, I) \leftarrow q(X).$$

and their negations, rather than using dual rules as obtained by Definition 5.2, can be defined to a single rule:

$$not\_q(X, I, I) \leftarrow not\ q(X).$$

Note that the input and output context arguments are added in the head, and the input context is just passed intact to the output context. Both rules simply execute the fact calls.

Facts of predicate $q/1$ can thus be defined in the so-called *non-abductive part* of the source program, independently of the number of facts $q/1$ are there in the program. The non-abductive part is distinguished from the abductive part by the beginProlog and endProlog identifiers. Any program between these identifiers will not be transformed, i.e., it is treated as a usual Prolog program. For the above facts of $q/1$, they are listed in the non-abductive part as:

$$\text{beginProlog} \quad q(1). \quad q(2). \quad q(3). \quad \text{endProlog}$$

5.1.2.4 Dual Transformation by Need

The TABDUAL transformation conceptually constructs *all* (first and second layer) dual rules, in advance, for every defined atom in an input program, regardless whether they are needed in abduction. We refer to this conceptual construction of dual rules in the sequel as the STANDARD dual program transformation.

The STANDARD dual program transformation should be avoided in practice, as potentially large sets of dual rules are created in the transformation, though only a few of them might be invoked during abduction. Consequently, it may burden the dual program transformation itself, affecting the time required to produce dual rules and the space required for the large thus produced transformed.

One solution to this problem is to compute dual rules only *by need*. That is, dual rules are concretely created in the abduction stage (rather than in the transformation stage), based on the need of the on-going invoked goals. The transformed program still contains the single first layer dual rule, but its second layer is defined using a newly introduced TABDUAL system predicate, which will be interpreted by the TABDUAL system on-the-fly, during the abduction stage, to produce the concrete rule definitions of the second layer.

Example 5.13 Recall Example 5.4. The dual transformation by need contains the same first layer: $not\_p(T_0, T_2) \leftarrow p^{*1}(T_0, T_1), p^{*2}(T_1, T_2)$. But the second now contains, for each $i \in \{1, 2\}$:

$$p^{*i}(I, O) \leftarrow dual(i, p, I, O). \tag{5.9}$$

Predicate $dual/4$ is a TABDUAL system predicate, introduced to facilitate the dual transformation by need:

1. It constructs *generic* dual rules, i.e., dual rules without specific context assigned to them, by need, from the ith rule of $p/0$. This construction is performed during abduction.
2. It instantiates the generic dual rules with the provided arguments and input context.
3. Finally, it subsequently invokes the instantiated dual rules.

The concrete definition of this predicate $dual/4$ depends on the dual rules construction mechanism detailed below.

While the dual program transformation by need minimizes the number of the second layer dual rules, constructing dual rules on-the-fly introduces some extra cost in the abduction stage. Such extra cost can be reduced by memoizing the already constructed generic dual rules. Therefore, when such dual rules are later needed, they are available for reuse and their recomputation avoided.

We examine two approaches for memoizing generic dual rules for the dual transformation by need. They lead to different definitions of the system predicate $dual/4$, particularly concerning how generic dual rules are constructed by need. The first approach benefits from tabling to memoize generic dual rules, whereas the second one employs the XSB Prolog's trie data structure [22]. They are referred in the sequel as BY- NEED(EAGER) and BY- NEED(LAZY), respectively, due to their dual rules construction mechanisms.

Dualization BY-NEED(EAGER): tabling generic dual rules

The straightforward choice for memoizing generic dual rules is to use tabling. The system predicate $dual/4$ in rule 5.9 is defined as follows:

$$dual(N, P, I, O) \leftarrow dual\_rule(N, P, Dual), call\_dual(P, I, O, Dual).$$

where $dual\_rule/3$ is a *tabled* predicate that constructs a generic dual rule $Dual$ from the Nth rule of atom P, and $call\_dual/4$ instantiates $Dual$ with the provided arguments of P and the input context I. It eventually invokes the instantiated dual rule to produce the abductive solution in O.

Though predicate *dual*/4 helps realize the construction of dual rules by need, i.e., only when a particular p^{*i} is invoked, this approach results in an *eager* construction of all dual rules for the *i*th rule of predicate *p*, when local table scheduling, like the one employed by default in XSB Prolog [21], is in place. This scheduling strategy does not return any answers out of a strongly connected component (SCC) in the subgoal dependency graph, until that SCC is completely evaluated [21].

As an illustration, in Example 5.4, when $p^{*2}(I, O)$ is invoked, which subsequently invokes *dual_rule*(2, *p*, *Dual*), all two alternatives of dual rules from the second rule of *p*, viz., $p^{*2}(I, O) \leftarrow not\_q(I, O)$ and $p^{*2}(I, O) \leftarrow r(I, O)$ are constructed before *call_dual*/4 is invoked for each of them. This is a bit against the spirit of a full dual transformation by need, whereby only one alternative dual rule is constructed at a time, just before it is invoked. That is, generic dual rules should rather be constructed *lazily*.

As an alternative to local table scheduling, batched scheduling is also implemented in XSB Prolog, which allows returning answers outside of a maximal SCC as they are derived. In terms of the dual rules construction by need, this means *dual_rule*/3 would allow dual rules to be lazily constructed. That is, only one generic dual rule is produced at a time before it is instantiated and invoked. Since the choice between the two scheduling strategies can only be made for the whole XSB Prolog installation, and is not (as yet) predicate switchable, we pursue another approach to implement lazy dual rule construction.

Dualization BY-NEED(LAZY): storing generic dual rules in a trie

Trie is a tree data structure that allows data, such as strings, to be compactly stored by a shared representation of their prefixes. That is, all the descendants of a node in a trie have a common prefix of the string associated with that node.

XSB Prolog offers a mechanism for facts to be directly stored and manipulated in tries. Figure 5.1, taken from [22], depicts a trie that stores a set of Prolog facts:

$$\{rt(a, f(a, b), a),\ rt(a, f(a, X), Y),\ rt(b, V, d)\}.$$

For trie-dynamic code, trie storage has advantages, both in terms of space and time [22]:

- A trie can use much less space to store many sets of facts than standard dynamic code, as there is no distinction between the index and the code itself.
- Directly inserting into or deleting from a trie is faster (up to 4–5 times) than with standard dynamic code, as discrimination can be made on a position anywhere in a fact.

XSB Prolog provides predicates for inserting terms into a trie, unifying a term with terms in a trie, and other trie manipulation predicates, both in the low-level and high-level interface.

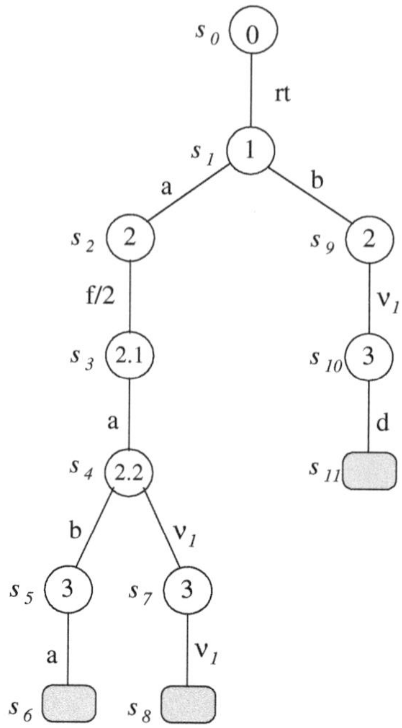

Fig. 5.1 Facts stored as a trie

Generic dual rules can be represented as facts, thus once they are constructed, they can be memoized and later (a copy) retrieved and reused. Given the aforementioned advantages for storing dynamic facts and XSB Prolog's support for its manipulation, a trie is preferable to the common Prolog database to store dynamically generated (i.e., by need) dual rules. The availability of XSB Prolog's system predicates to manipulate terms in a trie permits explicit control in lazily constructing generic dual rules compared to the more eager tabling approach, as detailed below.

A fact of the form $d(N, P, Dual, Pos)$ is used to represent a generic dual rule *Dual* from the *N*th rule of *P* with the additional tracking information *Pos*, which informs the position of the literal used in constructing each dual rule. For now, we opt for the low-level trie manipulation predicates, as they can be faster than the higher-level ones.

Using this approach, the system predicate *dual*/4 in rule 5.9 is defined as follows:

1. $dual(N, P, I, O) \leftarrow trie\_property(T, alias(dual)), dual(T, N, P, I, O).$

2. $dual(T, N, P, I, O) \leftarrow trie\_interned(d(N, P, Dual, \_), T),$
   $call\_dual(P, I, O, Dual).$
3. $dual(T, N, P, I, O) \leftarrow current\_pos(T, N, P, Pos),$
   $dualize(Pos, Dual, NextPos),$
   $store\_dual(T, N, P, Dual, NextPos),$
   $call\_dual(P, I, O, Dual).$

Assuming that a trie T with alias *dual* has been created, predicate $dual/4$ (line 1) is defined by an auxiliary predicate $dual/5$ with an access to the trie T, the access being provided by the trie manipulation predicate *trie_property*/2. Lines 2 and 3 give the definition of $dual/5$:

- In the first definition (line 2), an attempt is made to reuse generic dual rules, which are stored already as facts $d/4$ in trie T. This is accomplished by unifying terms in T with $d(N, P, Dual, \_)$, one at a time through backtracking, via the trie manipulation predicate *trie_interned*/2. Predicate *call_dual*/4 then does the job as before.
- The second definition (line 3) constructs generic dual rules lazily. It finds, via *current_pos*/4, the current position *Pos* of the literal from the Nth rule of P, which can be obtained from the last argument of fact $d(N, P, Dual, Pos)$ stored in trie T. Using this *Pos* information, a new generic dual rule *Dual* is constructed by means of *dualize*/3. The predicate *dualize*/3 additionally updates the position of the literal, *NextPos*, for the next dualization. The dual rule *Dual*, together with the tracking information, is then memoized as a fact $d(N, P, Dual, NextPos)$ in trie T, via *store_dual*/5. Finally, the just constructed dual *Dual* is instantiated and invoked using *call_dual*/4.

Whereas the first approach constructs generic dual rules by need eagerly, the second one does it lazily. But this requires memoizing dual rules to be carried out explicitly, and additional tracking information is needed to correctly pick up on dual rule generation at the point where it was last left. This approach affords us a simulation of batched table scheduling for *dual*/5, within the default local table scheduling.

5.1.3 Concluding Remarks

TABDUAL is an ongoing work, which primarily intends to sensitize a general audience of users, and of implementers of various LP systems, to the potential benefits of tabling in abduction. TABDUAL has been evaluated with various evaluation objectives [17]:

- We evaluate the benefit of tabling abductive solutions, where we employ an example from declarative debugging to debug missing solutions of logic programs, via a process now characterized as abduction [16], instead of as belief revision [9, 10].
- We use the other case of declarative debugging, that of debugging incorrect solutions, to evaluate the relative worth of the dual transformation by need.
- We touch upon tabling abductive solution candidates that violate constraints, the so-called *nogoods* of subproblems in the context of abduction, and show that tabling abductive solutions can be appropriate for this purpose.
- We also evaluate TABDUAL in dealing with programs having loops, where we compare its results with those obtained from an implementation of ABDUAL [5].

Other implementation aspects of TABDUAL, such as dealing with programs having loops and dynamically accessing ongoing abductive solutions for preferring amongst them, as well as the evaluation of TABDUAL, are discussed at length in [17]. They are beyond the scope of this book.

TABDUAL still has much room for improvement. Future work will consist in continued exploration of our applications of abduction, which will provide feedback for system improvement. Tabled abduction may benefit from answer subsumption [20] in tabling abductive solutions to deal with redundant explanations, in the sense that it suffices to table only smaller abductive solutions (with respect to the subset relation). Another potential XSB Prolog's feature to look into is the applicability of interning ground terms [23] for tabling abductive solutions, which are ground, and study how extra efficiency may be gained from it.

The implementation technique of BY- NEED(LAZY) consists in operational details that are facilitated by XSB Prolog's trie manipulation predicates, to simulate the batched-like table scheduling within XSB Prolog's current default local table scheduling. In order to have a more transparent implementation of those operations, it is desirable that XSB Prolog permit a mixture in using batched and local table scheduling strategies, or alternatively, stopping the evaluation at some first answers to a subgoal within the currently default local table scheduling.

Though TABDUAL is implemented in XSB Prolog, a number of its techniques are adaptable and importable into other LP systems that afford required tabling mechanisms. They add and aid to the considerations involved in the research of the still ongoing developments of tabling mechanisms in diverse LP systems, and serve to inspire these systems in terms of solutions, options and experimentation results of incorporating tabled abduction.

5.2 Incremental Tabling of Fluents for LP Updating

Incremental tabling [12, 19], available in XSB Prolog, is an advanced recent tabling feature that ensures the consistency of answers in a table with all dynamic clauses on which the table depends. It does so by incrementally maintaining the table, rather than by recomputing answers in the table from scratch to keep it updated. The applications of incremental tabling in LP have been demonstrated in pointer analyses of C programs in the context of incremental program analyses [13], data flow analyses [14], static analyses [7], incremental validation of XML documents and push down model checking [12]. This range of applications suggests that incremental tabling lends itself to dynamic environments and evolving systems, including notably logic program updating.

We conceptualize a technique with incremental tabling that permits a reconciliation of high-level top-down deliberative reasoning about a goal, with autonomous low-level bottom-up world reactivity to ongoing updates. The technique, dubbed EVOLP/R, is theoretically based on Dynamic Logic Programs [2] and its subsequent development, Evolving Logic Programs (EVOLP) [1].

5.2.1 The EVOLP/R Language

The syntax of EVOLP/R adopts that of *generalized logic programs*, following the definitions in [2], which is also the basis of the EVOLP language [1]. We leave out EVOLP's reserved predicate *assert*/1 in the EVOLP/R language, which is enough for our purpose in this book. Moreover, updates are restricted to fluents only, which is explained below.

Generalized logic programs allow negative information to be represented in logic programs and in their updates, due to the possibility to have a default negation not only in the body of a rule but also in its head. As in [2], for convenience, generalized logic programs are *syntactically* represented as propositional Horn theories. In particular, default negation *not A* is represented as a standard propositional atom *not_A*.

Let $\mathcal{K}$ be an arbitrary set of propositional atoms whose names do not begin with a "*not_*". The propositional language $\mathcal{L}_{\mathcal{K}}$ *generated* by $\mathcal{K}$ is the language whose set of propositional atoms consists of:

$$\{A : A \in \mathcal{K}\} \cup \{not\_A : A \in \mathcal{K}\}.$$

Definition 5.6 (*Generalized Logic Program*) A *generalized logic program* over the language $\mathcal{L}_{\mathcal{K}}$ is a countable set of rules of the form:

$$L \leftarrow L_1, \ldots, L_n$$

where L and L_is are atoms from $\mathcal{L}_{\mathcal{K}}$.

The evolution of a program is formed through a sequence of program updates. As in EVOLP, the sequence of programs are treated as in Dynamic Logic Programs (DLPs). We denote a set of *states* as $\mathcal{S} = \{1, 2, \ldots, s, \ldots\}$ of natural numbers. Let $\mathcal{P} = \{P_i \;:\; i \in \mathcal{S}\}$. A *dynamic logic program* $\bigoplus_s \mathcal{P}$ is a sequence of programs $P_1 \oplus \cdots \oplus P_s$. If the set S has the largest element *max*, we simply write $\bigoplus \mathcal{P}$ instead of $\bigoplus_{max} \mathcal{P}$.

In EVOLP/R, the evolution of a program P of the language $\mathcal{L}_{\mathcal{K}}$ from the initial state 1 to a state s is specified as a dynamic logic program $\bigoplus_s \mathcal{P}$, where:

- $P = P_1$ (the program P is given at the initial state 1).
- For $2 \leq i \leq s, P_i$ is a set of atoms from $\mathcal{L}_{\mathcal{K}}$, referred as *fluents* (i.e., state-dependent atoms).

The *negation complement of a fluent* A is denoted by $compl_{\mathrm{FL}}(A)$, where the complement of atom A and its negation *not_A* is defined as $compl_{\mathrm{FL}}(A) = not\_A$ and $compl_{\mathrm{FL}}(not\_A) = A$, respectively.

Next, we define the semantics of a DLP in EVOLP/R. As we consider a propositional language $\mathcal{L}_{\mathcal{K}}$, we first adapt Definition 4.2 of two-valued interpretations. Let $F \subseteq \mathcal{K}$ and *not* F represent $\{not\_A : A \in F\}$. A two-valued interpretation M over $\mathcal{L}_{\mathcal{K}}$ is a set of atoms $T \cup not\ F$ from $\mathcal{L}_{\mathcal{K}}$, such that $T \cup F = \mathcal{K}$ and $T \cap F = \emptyset$.

Definition 5.7 Let $\bigoplus\{P_i : i \in \mathcal{S}\}$ be a dynamic logic program over language $\mathcal{L}_{\mathcal{K}}$, $s \in S$, and M be a two-valued interpretation over $\mathcal{L}_{\mathcal{K}}$. Then:

$$Default_s(M) = \{not\_A \mid \nexists A \leftarrow Body \in P_i,\ 1 \leq i \leq s,\ M \models Body\}$$
$$Reject_s(M) = \{A \leftarrow Body \in P_i \mid \exists\ compl_{\mathrm{FL}}(A) \leftarrow Body' \in P_j, i < j \leq s\ \wedge M \models Body'\}$$

where both *Body* and *Body'* are conjunctions of literals.

Observe that since updates in EVOLP/R are restricted to fluents only, $M \models Body'$ in the definition of $Reject_s(M)$ is vacuously true, as we have an empty *Body'* for an updating fluent $compl_{\mathrm{FL}}(A)$.

Definition 5.8 A two-valued interpretation M over $\mathcal{L}_{\mathcal{K}}$ is a stable model of a dynamic logic program $\bigoplus\{P_i\ :\ i \in \mathcal{S}\}$ at state $s \in \mathcal{S}$ iff:

$$M = least\left(\left[\bigcup_{i \leq s} P_i - Reject_s(M)\right] \cup Default_s(M)\right)$$

The emphasis of EVOLP/R is on the implementation technique to demonstrate an innovative use of tabling, particularly the incremental tabling afforded by XSB Prolog, for dynamic program updating. Whereas XSB Prolog computes the Well-Founded Semantics, the semantics of dynamic logic programs in Definition 5.8 is based on the stable model semantics due to its simplicity.[2] Moreover, EVOLP/R currently considers only stratified programs (programs with no loops over negation), and the semantics for such programs therefore consists of only one stable model, which is also the well-founded model. This constraint is deliberately so made, at this point, as we want to concentrate on the incremental tabling aspects and usage for logic program updating, and later in its combination with abduction. The usage of incremental tabling for non-stratified programs within EVOLP/R, viz., for updating conditional answers, is a future line of work.

5.2.2 *Incremental Tabling*

Whenever a tabled predicate depends on dynamic predicates and the latter are updated (with Prolog's assert or retract predicates), these updates are not immediately reflected in the table, i.e., the table becomes out of date. This problem is known as the view maintenance problem in databases and the truth maintenance problem in artificial intelligence.

In "classical" tabling, a typical solution to this problem is to rely on the user to explicitly abolish the table whenever a dynamic predicate, on which the table depends, is updated. As several updates may take place on a dynamic predicate, such explicit table abolishment is rather inconvenient and defeats the benefit of tabling

[2] See [3] for the Well-Founded Semantics of generalized logic programs and [6] for the Well-Founded Semantics of dynamic logic programs.

itself, because those solutions in the abolished table have to be recomputed from scratch.

In order to overcome this problem, XSB Prolog allows maintaining particular tables incrementally, known as *incremental tabling*. That is, the answers in these tables are ensured to be consistent with all dynamic facts and rules upon which they depend. In XSB Prolog, this requires both tabled predicates and the dynamic predicates they depend on to be declared as incremental. By default, the current XSB Prolog (version 3.6) updates an incremental table *transparently* [22]: after a sequence of updates Δ, an incremental table T that depends on Δ and all tables upon which T depends are automatically updated (if needed) whenever a future subgoal calls T. This transparent incremental tabling promotes a *lazy* updating strategy, where table T is marked as *invalid*, and when a subgoal calls this invalid table T (after the invalidation phase is completed), then T and any tables upon which T depends are recomputed to reflect the updates. On the other hand, if no calls are ever made to an invalid incremental table, it will never incur the cost of an update. This is in contrast with the eager updating strategy of the earlier XSB Prolog's version, in which invalidated tables are updated immediately.

Example 5.14 below demonstrates how incremental tabling is used.

Example 5.14 We first consider the following program that does not use incremental tabling:

$$r(X, Y) \leftarrow s(X, Y), Y < 4.$$
$$s(a, 2).$$
$$s(b, 4).$$

where $r/2$ is a tabled predicate, declared in XSB Prolog as

```
:- table r/2.
```

and $s/2$ is a dynamic predicate, declared as

```
:- dynamic s/2.
```

Query `?-` r(X,Y) succeeds with $X = a$ and $Y = 2$. Suppose a new fact $s(c, 1)$ is asserted, via a usual Prolog assertion `assert(s(c,1))`. The answer of the same query `?-` r(X,Y) has not changed. The new solution $X = c$ and $Y = 1$ is not part of the query's solution, because the table is already created and the second invocation of the query just retrieves the existing single answer directly from the table.

With incremental tabling, XSB Prolog automatically keeps the table for $r/2$ correct with respect to the given update, returning also the new answer $X = c,\ Y = 1$. This is achieved by declaring $r/2$ as an incremental tabled predicate:

```
:- table r/2 as incremental.
```

and $s/2$ as an incremental dynamic predicate:

```
:- dynamic s/2 as incremental.
```

Moreover, a specific incremental assertion `incr_assert(s(c,1))` should be used for incremental tabling instead of a usual `assert(s(c,1))`.

The reader is referred to [22] for the further examples, predicates, and features of incremental tabling.

5.2.3 *The* EVOLP/R *Approach*

Since incremental tabling allows tables to be correctly updated according to changes made by incremental assertions (or retractions), possibly in a chain of dependencies between tabled predicates, this feature can be exploited for automatically propagating fluent updates while tabling the states that indicate when a fluent holds true. By tabling this state information the recursive nature of the inertia principle can be avoided, as tabling readily provides the history of the world, and incremental tabling maintains this history to reflect changes made to the world.

The concept of the EVOLP/R approach is depicted in Fig. 5.2 and is summarized below:

1. A program transformation is first applied to transform the input program into the processed one. The transformation is responsible for, amongst others, adding extra information for rules (e.g., unique rule names, information about states, incremental table and dynamic declaration), providing an interface to EVOLP/R system predicates, and adding dual rules (similar to those used in TABDUAL,

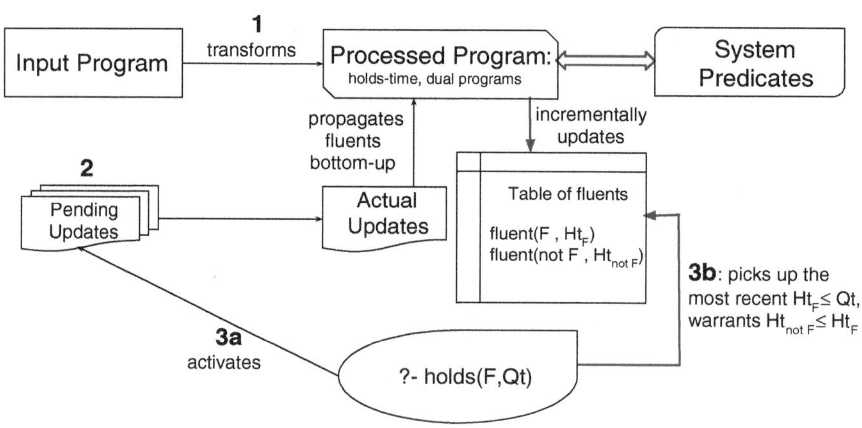

Fig. 5.2 The concept of the EVOLP/R approach

see Definition 5.2) to help propagate the dual negation complement of a fluent incrementally, to be explained below.

2. The issued fluent updates are initially kept pending in the database.
3. Query ?- $holds(F, Qt)$ is a top-goal query to test whether fluent F holds true at state Qt (referred to as *query-time*).

    a. On the initiative of this query, i.e., by need only, incremental assertions make these pending updates become active, if not already so, but only those with states up to an actual Qt. Such assertions automatically trigger Prolog system-implemented incremental bottom-up propagation of fluent updates, recomputation of answers for affected tables.
    b. After the table is correctly maintained, this query is answered by looking up a collection of states fluent F is true in the table, picking up the most recent one, and ensuring that its negation complement *not_F* with a later state does not exist in the table.

5.2.3.1 The EVOLP/R Transformation

We start with a simple example that illustrates what is required from the EVOLP/R transformation.

Example 5.15 Let $\mathcal{P} = \{P_1, P_2, P_3\}$, where:

$$\begin{aligned} P_1 &: b \leftarrow a. \\ &\quad c \leftarrow b. \\ P_2 &: a. \\ P_3 &: not\_a. \end{aligned}$$

Program P_1 evolves from the initial state 1 through a series of fluent updates: it is updated at state 2 with fluent a, and at state 3 with fluent not_a. We introduce an incremental tabled predicate $\mathit{fluent}(F, T)$ to incrementally record that fluent F is true at a state T. The update of fluent a in P_2 (at state $i = 2$) is accomplished via the XSB Prolog's $incr\_assert/1$ system predicate, viz., by the incremental assertion $incr\_assert(a(2))$ to say that fluent a is incrementally asserted at state $i = 2$. Such an incremental assertion results in having entry $\mathit{fluent}(a, 2)$ in the table. Due to the dependencies of the three fluents, as defined by the two rules in P_1, the incremental assertion of a propagates to fluents b and c, leading to tabling $\mathit{fluent}(b, 2)$ and $\mathit{fluent}(c, 2)$. We thus have $\mathit{fluent}(a, 2)$, $\mathit{fluent}(b, 2)$, and $\mathit{fluent}(c, 2)$ that the three fluents are true at state $i = 2$, which conforms with the stable model $M_2 = \{a, b, c\}$ of $\bigoplus_2 \mathcal{P}$.

A subsequent update of fluent not_a at state $i = 3$ via $incr\_assert(not\_a(3))$ results in tabling $\mathit{fluent}(not\_a, 3)$. That means, we still have all previous tabled entries, viz., $\mathit{fluent}(a, 2)$, $\mathit{fluent}(b, 2)$, and $\mathit{fluent}(c, 2)$, plus now $\mathit{fluent}(not\_a, 3)$. While a is no longer true at this state ($i = 3$), which now can easily be checked by comparing states of a in the table (fluent a at $i = 2$ is supervened by its complement *not_a* at a later state $i = 3$), the same reasoning cannot be applied to fluents b and c. Different from before, the incremental assertion of *not_a* unfortunately does not propagate to tabling fluents *not_b* an *not_c*, i.e., neither *fluent*(*not_b*, 3) nor *fluent*(*not_c*, 3) found in the table. Indeed, there are no corresponding rules in P for *not_b* and *not_c*; thus failing to conclude that both fluents are also false at $i = 3$ by propagating *not_a* (cf. $M_3 = \{not\_a, not\_b, not\_c\}$ of $\bigoplus_3 \mathcal{P}$). In order to provide rules for *not_b* and *not_c*, we adopt the dual program transformation, which is similar to that we use in TABDUAL, see Definition 5.2, but shown below by unfolding the first and second layers dual rules:

$$not\_b \leftarrow not\_a.$$

$$not\_c \leftarrow not\_b.$$

The introduced dual rules now allow the propagation from *not_a* to *not_b* and then to *not_c*, resulting in having *fluent*(*not_b*, 3) and *fluent*(*not_c*, 3) in the table. By having the latter two entries in the table, using the same previous reasoning, it can be concluded that fluents b and c are also false at $i = 3$, confirming M_3 of $\bigoplus_3 \mathcal{P}$.

This example hints at important information to be added in the transformation:

1. *Timestamp* that corresponds to state and serves as the only extra argument of fluents. It denotes the state when a fluent becomes true, referred to as *holds-time*.
2. *Dual rules* that are obtained using the transformation based on Definition 5.2.

The transformation technique is illustrated by Example 5.16 below, and is explained subsequently.

Example 5.16 Recall program P_1:

$$b \leftarrow a.$$

$$c \leftarrow b.$$

which transforms into:

1. $\#r(b, [a], 1).$
2. $\#r(c, [b], 1).$
3. $b(H) \leftarrow fluent(\#r(b, [a]), H_r),\ fluent(a, H_a),\ latest([(\#r(b, [a]), H_r), (a, H_a)], H).$
4. $not\_b(H) \leftarrow fluent(not\_\#r(b, [a]), H).$
5. $not\_b(H) \leftarrow fluent(not\_a, H).$
6. $c(H) \leftarrow fluent(\#r(c, [b]), H_r),\ fluent(b, H_b),\ latest([(\#r(c, [b]), H_r), (b, H_b)], H).$
7. $not\_c(H) \leftarrow fluent(not\_\#r(c, [b]), H).$
8. $not\_c(H) \leftarrow fluent(not\_b, H).$
9. $not\_a(1).$

In EVOLP/R, the initial timestamp is set at 1, when a program is inserted. Fluent predicates can be defined as facts (extensional) or by rules (intensional). In Example 5.16, both fluents b and c are defined intensionally. For such rule regulated intensional fluent instances, unique rule name fluents are introduced. A *rule name* fluent is a special fluent $\#r(H, [B])$, which uniquely identifies the rule $H \leftarrow B$, and is introduced in its body, for controlling the activation of the rule (see [11]). In Example 5.16, rule name fluents $\#r(b, [a])$ and $\#r(c, [b])$ are introduced for rules $b \leftarrow a$ and $c \leftarrow b$, respectively. They are extensional fluent instances, and like any other extensional fluent instances, such a rule name fluent is translated by adding an extra argument (the third one) that corresponds to its holds-time (lines 1 and 2). In this case, each rule name fluent is true at the initial time 1, viz., the time when its corresponding rule is inserted.

Line 3 shows the translation of rule $b \leftarrow a$ of the input program. The single extra argument in its head is its holds-time, H. The body, which originally comprises just a call to goal a, is translated into two calls wrapped with the EVOLP/R's reserved *incremental tabled* predicate *fluent*/2 (defined in Sect. 5.2.3.2), viz., $\mathit{fluent}(\#r(b, [a]), H_r)$ and $\mathit{fluent}(a, H_a)$, which provides holds-time H_r and H_a of fluents $\#r(b, [a])$ and a, respectively. Note the inclusion of the unique rule name fluent (i.e., the call $\mathit{fluent}(\#r(b, [a]), H_r)$) in the body, whose purpose is to switch the corresponding rule on or off; this being achieved by asserting the rule name fluent or its negation complement, respectively. The holds time H of fluent b in the head is thus determined by which inertial fluent in its body holds the latest, via the EVOLP/R's *latest*/2 reserved predicate (defined in Sect. 5.2.3.2), which also assures that no fluents in the body were subsequently supervened by their complements at some time before H.

Lines 4 and 5 show the dual rules for b in their flattened form (unfolding the first and second layers dual rules). Line 4 expresses how the negation complement $not\_\#r(b, [a])$ of rule name fluent $\#r(b, [a])$ propagates to fluent not_b, whereas line 5 expresses the other alternative: how the negation complement not_a of fluent a propagates to fluent not_b.

Similar technique is applied to rule $c \leftarrow b$, resulting in rules shown in lines 6–8. Finally, line 9 is the dual rule for atom a that has no rule in P_1.

Since every fluent occurring in the program is subject to updates, all fluents and their negation complements should be declared as dynamic and incremental; the latter attribute is due to incremental tabling of fluents by the incremental tabled predicate *fluent*/2. For Example 5.16, we have, e.g.,

```
:- dynamic a/1, not_a/1 as incremental.
```

and similarly for fluents b, c, $\#r(b, [a])$, $\#r(c, [b])$, as well as their negation complements.

5.2.3.2 Reserved Predicates

Predicate *fluent*(F, T) used in the transformation is a tabled one. Its dependency on fluent F, which is dynamic incremental, indicates that *fluent*/2 is tabled *incrementally*, and therefore declared

```
:- table fluent/2 as incremental.
```

This predicate is and defined as follows:

$$fluent(F, T) \leftarrow extend(F, [T], F'),\ call(F').$$

where $extend(F, Args, F')$ extends the arguments of fluent F with those in list $Args$ to obtain F'.

Updates propagation in EVOLP/R is query-driven, within some query-time of interest. This means we can use the given query-time to control updates propagation by keeping the sequence of updates pending in the database, and then making active, through incremental assertions, only those with the states up to the actual query-time (if they have not yet been so made already by queries of a later time stamp).

For expressing pending updates, we introduce a dynamic predicate $pending(F, T)$ to indicate that update of fluent F at state T is still pending, and use Prolog *assert*/1 predicate, i.e., *assert*(*pending*(F, T)), to assert such a pending fluent update into the Prolog database. Activating pending updates (up to the actual query-time Qt), as shown by the code below, can thus be done by calling all $pending(F, T)$ facts with $T \leq Qt$ from the database and actually asserting them incrementally using the XSB Prolog's system predicate *incr_assert*/1:

$$\begin{aligned} activate\_pending(Qt) \leftarrow\ & pending(F, T),\ T \leq Qt,\ extend(F, [T], F'),\\ & incr\_assert(F'),\ retract(pending(F, T)), fail.\\ activate\_pending(\_).\ & \end{aligned}$$

We have seen predicate $latest([(F_1, H_1), \ldots, (F_n, H_n)], H)$ in the transformation, which appears in the body of a rule transform, say of fluent F. This reserved predicate is responsible for obtaining the latest holds-time H of F amongst fluents $F_1, \ldots, F_n$ in the body, while also assuring that none of them were subsequently supervened by their complements at some time up to H. It is defined as:

$$latest(Fs, H) \leftarrow greatest(Fs, H),\ not\_supervened(Fs, H).$$

where $greatest(Fs, H)$ extracts from list Fs, of (F_i, H_i) pairs with $1 \leq i \leq n$, the greatest holds-time H among the Hi's. The predicate $not\_supervened(Fs, H)$ subsequently guarantees that there is no fluent complement F_i' (with holds time H_i') of F_i in Fs, such that $H_i < H_i' \leq H$.

Finally, the top-goal query $holds(F, Qt)$ in EVOLP/R (see Fig. 5.2) is defined below. This reserved predicate is introduced to test whether fluent F is true at query-time Qt by first activating pending updates up to Qt:

$$\begin{aligned} holds(F, Qt) \leftarrow\ & activate\_pending(Qt),\ compl(F, F'), \\ & most\_recent(F, H_F, Qt),\ most\_recent(F', H_{F'}, Qt), \\ & H_F \neq 0,\ H_F \geq H_{F'}. \end{aligned}$$

where the predicate $compl(F, F')$ obtains the fluent negation complement $F' = compl_{\text{FL}}(F)$, and the predicate $most\_recent(F, H_F, Qt)$ calls $fluent(F, H)$ and picks up the entry of fluent F from the table with the highest timestamp $H_F \leq Qt$ (or returns $H_F = 0$, if the call $fluent(F, H)$ fails). For fluent F to hold true in the query-time Qt, this definition warrants that F ($H_F \neq 0$, for F not to trivially fail) is not supervened by its complement F', i.e., $H_F \geq H_{F'}$, where H' is obtained by invoking $most\_recent(F', H_{F'}, Qt)$.

5.2.4 Concluding Remarks

The EVOLP/R approach is somewhat similar and complementary approach to Logic-based Production System (LPS) with abduction [8]. The latter aims at defining a new logic-based framework for knowledge representation and reasoning, relying on the fundamental role of state transition systems in computing, and involving fluent updates by *destructive* assignment. In EVOLP/R, fluent updates are not managed by destructive database assignments, but rather tabled, thereby allowing to inspect their truths at a particular time, e.g., querying the past, which is important in counterfactual reasoning, as we discuss in Chap. 6.

Our first approach of EVOLP/R [15] preliminarily exploits the combination of incremental tabling and *answer subsumption* [20]. The latter allows tables to retain only answers that subsume others with respect to some order relation. In that first approach, answer subsumption of fluent literals aims at addressing the frame problem, i.e., by automatically keeping track of only their *latest* assertion with respect to a given query-time.

The combined use of incremental tabling and answer subsumption is realized in the incrementally tabled predicate $fluent(F, Ht_F, Qt)$ for fluent literal F, where Ht_F and Qt are the holds-time of F and the query-time, respectively. Note the extra argument Qt in this specification of $fluent/3$. Invoking $fluent(F, Ht_F, Qt)$ thus, either looks for an entry in its table, if one exists; otherwise, it invokes dynamic definitions of fluent F, and returns *the* latest holds-time Ht_F with respect to a given query-time Qt. In order to return only the latest holds-time Ht_F (with respect to Qt), $fluent/3$ is tabled using answer subsumption on its second parameter, which is declared in XSB Prolog as:

```
:- table fluent(_,po('>'/2),_) as incremental.
```

to mean that only those answers that are maximal according to the partial order $>/2$ (arithmetic greater-than comparison relation) are tabled. In terms of *fluent*/3, it returns the latest holds-time (the second parameter in which the answer subsumption is applied) within a given query-time.

While answer subsumption is shown useful in this approach to avoid recursing through the frame axiom by allowing direct access to the latest time when a fluent is true, it requires *fluent*/3 to have query time Qt as its argument. Consequently, it may hinder the reusing of tabled answers of *fluent*/3 by similar goals which differ only in their query-time. Ideally, the state of a fluent literal in time depends solely on the changes made to the world, and not on whether that world is being queried. As an illustration, suppose $fluent(a, 2, 4)$ is already tabled, and fluent a is inertially true till it is supervened by its negation complement not_a, say at time $T = 7$. When a new goal $fluent(a, Ht, 5)$ is posed, it cannot reuse the tabled answer $fluent(a, 2, 4)$, as they differ in their query time: the latter with $Qt = 4$, whereas the former $Qt = 5$. In this case, $fluent(a, Ht, 5)$ unnecessarily recomputes the same solution $Ht = 2$ (as fluent a is only supervened at $T = 7 > 5 = Qt$) and subsequently tables $fluent(a, 2, 5)$ as a new answer. A similar situation occurs when $fluent(a, Ht, 6)$ is queried, where $fluent(a, 2, 6)$ is eventually added into the table. This is clearly superfluous, as existing tabled answers could actually be reused and such redundancies avoided, if the tabled answers are independent of query time.

The above issue is addressed by our approach detailed in Sect. 5.2, where the use of incremental tabling in EVOLP/R is fostered further, while leaving out the problematic use of answer subsumption. The main idea, not captured in the first approach, is the perspective that knowledge updates (either self or world wrought changes) occur whether or not they are queried: the former take place independently of the latter, i.e., when a fluent is true at Ht, its truth lingers on independently of Qt. Consequently, from the standpoint of the tabled *fluent* predicate definition, Qt no longer becomes its argument: we now have incremental tabled predicate $fluent(F, Ht)$.

The current EVOLP/R approach can also be extended by considering the reserved predicate *assert*/1 (not to be confused with the Prolog predicate *assert*/1) into its language, as introduced in EVOLP [1]. By having $assert(F)$ in the head of a rule, the program is updated by fluent F, whenever the assertion $assert(F)$ is true in a model; or retracts F in case $assert(not\_F)$ obtains in the model under consideration. However, the EVOLP/R transformation becomes more involved. For instance, consider the rule:

$$assert(a) \leftarrow b.$$

The transformation of this rule is illustrated below:

$$
\begin{array}{lll}
1. & assert(a,H) & \leftarrow fluent(\#r(assert(a),[b]),H_r),\ fluent(b,H_b),\\
 & & \quad latest([(\#r(assert(a),[b]),H_r),(b,H_b)],H).\\
2. & a(H) & \leftarrow fluent(assert(a),H_{as}),\ H \text{ is } H_{as}+1.\\
3. & not\_assert(a,H) & \leftarrow fluent(not\_\#r(assert(a),[b]),H).\\
4. & not\_assert(a,H) & \leftarrow fluent(not\_b,H).\\
5. & not\_b(1). &
\end{array}
$$

This rule transforms into two rules. The rule in line 1 is obtained following the same transformation as before, by treating *assert*(a) as a fluent. The rule in line 2 is derived as the effect of asserting a. That is, the truth of a is determined solely by the propagation of fluent *assert*(a), indicated by the call $fluent(assert(a), H_{as})$. The holds time H of a is thus determined by $H_{as} + 1$, instead of H_{as}, because a is actually asserted one state ahead after the state at which *assert*(a) holds.

The other rules in lines 3–5 are dual rules for fluent *assert*(a) obtained similarly as before. Note that rule in line 2 does not produce any dual rule. From the semantics viewpoint, once a is asserted, its truth remains intact by inertia till superseded, even if *assert*(a) is retracted at a later time.

Introducing *assert*/1 into the language may require EVOLP/R to take care of non-termination of updates propagation. Consider program P below:

$$assert(not\_a) \leftarrow a.$$
$$assert(a) \leftarrow not\_a.$$

where a is true in a state s when a is asserted into the program, *not_a* is true in state $(s + 1)$ as *not_a* is asserted subsequently, a is true again in state $(s + 2)$, etc.; the evolution continues indefinitely. From the incremental tabling viewpoint, it indicates that a predefined upper time limit is required to delimit updates propagation, thereby avoiding infinite number of answers in the *fluent*/2 table. This requirement is realistic, as our view into the future may be bounded by some time horizon, analogous to bounded rationality. Such delimitation can be done via a predicate, say *upper_time*(*Lim*), indicating the predefined upper time limit *Lim*, and is called in the definition of *fluent*/2 to time-delimit their tabled answers, modified as shown below:

$$fluent(F,T) \leftarrow upper\_time(Lim),\ extend(F,[T],F'),\ call(F'),\ T \leq Lim.$$

In this example, when the fact *upper_time*(4) is given, the *fluent*/2 table will contain a finite number of answers concerning fluent a and its negation complement *not_a*, viz., $fluent(not\_a, 1)$, $fluent(a, 2)$, $fluent(not\_a, 3)$, and $fluent(a, 4)$.

The extension of the present EVOLP/R approach to deal with this assertion construct, and its issues that may arise from the use of incremental tabling, are to be explored in future.

Given our need to equip abduction with updating and that our TABDUAL technique affords tabling, the exploitation of incremental tabling for LP updating, as in EVOLP/R, anticipates the integration between the two reasoning features. To this end, we want to ensure that updates occurring in abduction should correctly maintain the table of abductive solutions incrementally, and incremental tabling provides this mechanism. As abduction in TABDUAL is accomplished by a top-down query-oriented procedure and the incremental tabling of fluents in EVOLP/R triggers bottom-up updates propagation, their integration thus permits top-down (deliberative) abduction to meet bottom-up (reactive) updates propagation, via tabling. We shall discuss this integration of LP abduction and updating with tabling in Sect. 7.3 of Chap. 7.

References

1. Alferes, J.J., Brogi, A., Leite, J.A., Pereira, L.M.: Evolving logic programs. In: Proceedings of the European Conference on Artificial Intelligence (JELIA 2002), LNCS, vol. 2424, pp. 50–61. Springer (2002)
2. Alferes, J.J., Leite, J.A., Pereira, L.M., Przymusinska, H., Przymusinski, T.: Dynamic updates of non-monotonic knowledge bases. J. Log. Progr. **45**(1–3), 43–70 (2000)
3. Alferes, J.J., Pereira, L.M., Przymusinski, T., Przymusinska, H., Quaresma, P.: Preliminary exploration on actions as updates. In: Proceeding of the Joint Conference on Declarative Programming (AGP 1999) (1999)
4. Alferes, J.J., Pereira, L.M., Swift, T.: Abduction in well-founded semantics and generalized stable models via tabled dual programs. Theory Pract. Log. Progr. **4**(4), 383–428 (2004)
5. Alferes, J.J., Pereira, L.M., Swift, T.: Abdual meta-interpreter. Available from http://www.cs.sunysb.edu/~tswift/interpreters.html (2004)
6. Banti, F., Alferes, J.J., Brogi, A.: Well founded semantics for logic program updates. In: Proceeding of the 9th Ibero-American Conference on Artificial Intelligence (IBERAMIA), LNCS, vol. 3315, pp. 397–407 (2004)
7. Eichberg, M., Kahl, M., Saha, D., Mezini, M., Ostermann, K.: Automatic incrementalization of prolog based static analyses. In: Proceedings of the 9th International Symposium on Practical Aspects of Declarative Languages (PADL), LNCS, vol. 4354, pp. 109–123. Springer (2007)
8. Kowalski, R., Sadri, F.: Abductive logic programming agents with destructive databases. Ann. Math. Artif. Intell. **62**(1), 129–158 (2011)
9. Pereira, L.M., Damásio, C.V., Alferes, J.J.: Debugging by diagnosing assumptions. In: Automatic Algorithmic Debugging, LNCS, vol. 749, pp. 58–74. Springer (1993)
10. Pereira, L.M., Damásio, C.V., Alferes, J.J.: Diagnosis and debugging as contradiction removal in logic programs. In: Progress in Artificial Intelligence, LNAI, vol. 727, pp. 183–197. Springer (1993)
11. Poole, D.L.: A logical framework for default reasoning. Artif. Intell. **36**(1), 27–47 (1988)
12. Saha, D.: Incremental evaluation of tabled logic programs. Ph.D. thesis, SUNY Stony Brook (2006)
13. Saha, D., Ramakrishnan, C.R.: Incremental and demand-driven points-to analysis using logic programming. In: Proceedings of the 7th ACM-SIGPLAN International Symposium on Principles and Practice of Declarative Programming (PPDP), pp. 117–128. ACM (2005)
14. Saha, D., Ramakrishnan, C.R.: A local algorithm for incremental evaluation of tabled logic programs. In: Proceedings of the 22nd International Conference on Logic Programming (ICLP), LNCS, vol. 4079, pp. 56–71. Springer (2006)

15. Saptawijaya, A., Pereira, L.M.: Program updating by incremental and answer subsumption tabling. In: Proceedings of the 12th International Conference on Logic Programming and Nonmonotonic Reasoning (LPNMR), LNCS, vol. 8148, pp. 479–484. Springer (2013)
16. Saptawijaya, A., Pereira, L.M.: Towards practical tabled abduction usable in decision making. In: Proceedings of the 5th KES International Conference on Intelligent Decision Technologies (IDT), Frontiers of Artificial Intelligence and Applications (FAIA). IOS Press (2013)
17. Saptawijaya, A., Pereira, L.M.: TABDUAL: a tabled abduction system for logic programs. IfCoLog J. Log. Appl. **2**(1), 69–123 (2015)
18. Swift, T.: Tabling for non-monotonic programming. Ann. Math. Artif. Intell. **25**(3–4), 201–240 (1999)
19. Swift, T.: Incremental tabling in support of knowledge representation and resoning. Theory Pract. Log. Progr. **14**(4–5), 553–567 (2014)
20. Swift, T., Warren, D.S.: Tabling with answer subsumption: Implementation, applications and performance. In: JELIA 2010, LNCS, vol. 6341, pp. 300–312. Springer (2010)
21. Swift, T., Warren, D.S.: XSB: extending Prolog with tabled logic programming. Theory Pract. Log. Progr. **12**(1–2), 157–187 (2012)
22. Swift, T., Warren, D.S., Sagonas, K., Freire, J., Rao, P., Cui, B., Johnson, E., de Castro, L., Marques, R.F., Saha, D., Dawson, S., Kifer, M.: The XSB System Version 3.6.x Volume 1: Programmer's Manual (2015)
23. Warren, D.: Interning ground terms in XSB. In: Colloquium on Implementation of Constraint and Logic Programming Systems (CICLOPS 2013) (2013)

Chapter 6
Counterfactuals in Logic Programming

Counterfactuals capture the process of reasoning about a past event that did not occur, namely what would have happened had this event occurred; or, vice-versa, to reason about an event that did occur but what if it had not. An example, taken from [4]: *Lightning hits a forest and a devastating forest fire breaks out. The forest was dry after a long hot summer and many acres were destroyed.* One may think of a counterfactual about it, e.g., "if only there had not been lightning, then the forest fire would not have occurred".

In this chapter, we counterfactual make use of LP abduction and updating in an implemented procedure for evaluating counterfactuals, taking the established approach of Pearl [19] as reference. Our approach concentrates on pure non-probabilistic counterfactual reasoning in LP, resorting to abduction and updating, in order to determine the logical validity of counterfactuals under the Well-Founded Semantics. Nevertheless, the approach is adaptable to other semantics, e.g., Weak Completion Semantics [13] is employed in [22].[1] Though abstaining from probability, this approach may be suitable and applicable to instances when probabilities are not known or needed. Abstaining from probability permits focusing on the naturalized logic of human counterfactual moral reasoning, as discussed in Sect. 3.3. In particular, morality aims at definitive (not probable) conclusions of right and wrong. Moreover, people naturally do not compute formal probabilities, nor probabilities are always available, when making moral decisions via counterfactuals, though one can benefit from counterfactuals for inferring intentions through a probabilistic model to explain moral permissibility [14]. Note that, even though the LP technique

[1]Both the Well-Founded Semantics (WFS) and the Weak Completion Semantics (WCS) are 3-valued semantics that differ in dealing with close world assumption (CWA) and rules with positive loops (e.g., $p \leftarrow p$). WFS enforces CWA, i.e., atom a that has no rule is interpreted as false, whereas in WCS undefined. Nevertheless, they can be transformed one to another: adding rules $a \leftarrow u$ and $u \leftarrow not\ u$ for a reserved atom u renders a unknown in WFS; alternatively, adding $a \leftarrow \bot$, where $\bot$ is false, enforces CWA in WCS. In this book, positive loops are not needed and do not appear throughout examples we consider.

L.M. Pereira and A. Saptawijaya, *Programming Machine Ethics*,
Studies in Applied Philosophy, Epistemology and Rational Ethics 26,
DOI 10.1007/978-3-319-29354-7_6

introduced in this chapter is relevant for modeling counterfactual moral reasoning, its use is general, not specific to morality.

Counterfactuals have been widely studied in philosophy [5, 11, 15], psychology [4, 8, 16–18, 23], as well as from the computational viewpoint [3, 9, 19, 20, 24]. In [19], counterfactuals are evaluated based on a probabilistic causal model and a calculus of intervention. Its main idea is to infer background circumstances that are conditional on current evidences, and subsequently to make a minimal required intervention in the current causal model, so as to comply with the antecedent condition of the counterfactual. The modified model serves as the basis for computing the counterfactual consequence's probability.

Instead of defining a new formalism for counterfactual reasoning, we adopt here Pearl's approach as an inspiration, but abstaining from probabilities—given the lack of pure non-probabilistic counterfactual reasoning in LP—by resorting to LP abduction and updating. LP lends itself to Pearl's causal model of counterfactuals. For one, the inferential arrow in a LP rule is adept at expressing causal direction. For another, LP is enriched with functionalities, such as abduction and defeasible reasoning with updates. They can be exploited to establish a LP non-probabilistic reconstruction of Pearl's counterfactuals evaluation procedure. That is, LP abduction is employed for discovering background conditions from observations made or evidences given, whereas defeasible logic rules allow achieving adjustments to the current model via hypothetical updates of intervention on the causal model.

We start, in Sect. 6.1, with a summary of Pearl's structure-based counterfactuals and how its main ingredients, viz., causation and intervention, can be captured in LP. We detail subsequently, in Sect. 6.2, our LP-based procedure to evaluate counterfactuals, and provide concluding remarks in Sect. 6.3.

6.1 Causation and Intervention in LP

Pearl [19] proposes a structural theory of counterfactuals based on a probabilistic causal model (likened to a Causal Bayesian Network) and a calculus of intervention (viz., his do-calculus). A causal model M consists of two sets of variables U and V, and a set F of functions that decides how values are assigned to each variable $V_i \in V$. The variables in U are background knowledge that have no explanatory mechanism encoded in model M. The values of all variables in V are uniquely determined by every instantiation $U = u$ of the background knowledge.

Procedure 1 *Given evidence e, the probability of the counterfactual sentence "Y would be y had X been x" can be evaluated in a three-step process:*

1. ***Abduction**: Update the probability $P(u)$ by the evidence e to obtain $P(u \mid e)$. This step explains the past circumstance $U = u$ in the presence of evidence e.*
2. ***Action**: Modify M by the action $do(X = x)$. This step minimally adjusts model M by a hypothetical intervention via the external action $do(X = x)$ to comply with the antecedent condition of the counterfactual.*

3. ***Prediction***: *Compute the probability $Y = y$ in the modified model. In this step the consequence of the counterfactual is predicted based on the evidential understanding of the past (Step 1), and the hypothetical modification performed in Step 2.*

In summary, the approach determines the probability of the counterfactual's consequence $Y = y$ by performing an intervention to impose the counterfactual's antecedent $X = x$ (other things being equal), given evidence e about $U = u$.

Two important constructions in Pearl's approach of counterfactuals are causal model and intervention. *Causation* denotes a specific relation of cause and effect. Causation can be captured by LP rules, where the inferential arrow in a logic rule represents causal direction. LP abduction is thus appropriate for inferring causation, providing explanation to a given observation. That said, LP abduction is not immediately sufficient for counterfactuals. Consider a simple logic program $P = \{b \leftarrow a\}$. Whereas abduction permits obtaining explanation a to observation b, the evaluation of counterfactual "if a had not been true, then b would not have been true" cannot immediately be evaluated from the conditional rule $b \leftarrow a$, for if its antecedent is false the counterfactual would be trivially true. That justifies the need for an *intervention*. That is, it requires explicitly imposing the desired truth value of a, and subsequently checking whether the predicted truth value of b consistently follows from this intervention. As described in Pearl's approach, such an intervention establishes a required adjustment, so as to ensure that the counterfactual's antecedent be met. It permits the value of the antecedent to differ from its actual one, whilst maintaining the consistency of the modified model. We resort to LP abduction and updating to express causal source and intervention, respectively.

6.1.1 Causal Model and LP Abduction

With respect to an abductive framework $\langle P, \mathcal{AB}, \mathcal{IC}\rangle$, observation O corresponds to Pearl's definition for evidence e. That is, O has rules concluding it in program P, and hence does not belong to the set $\mathcal{AB}_{gL}$. Recall from Sect. 4.2 that $\mathcal{AB}_{gL}$ refers to the set of ground abducibles formed over the set of abducible predicates $\mathcal{AB}$.

In Pearl's approach, a model M consists of set U of background variables, whose values are conditional on case-considered observed evidences. These background variables are not causally explained in M, as they have no parent nodes in the causal diagram of M. In terms of LP abduction, they correspond to a set of abducibles $E \subseteq \mathcal{AB}_{gL}$ that provide abductive explanations to observation O. Indeed, these abducibles likewise have no preceding causal explanatory mechanism, as they have no rules concluding them in the program.

In a nutshell, an abductive framework $\langle P, \mathcal{AB}, \mathcal{IC}\rangle$ that provides an abduced explanation $E \subseteq \mathcal{AB}_{gL}$ to the available observation O mirrors Pearl's model M with its specific U supporting an explanation to the current observed evidence e.

6.1.2 Intervention and LP Updating

Besides abduction, our approach benefits from LP updating, which allows a program to be updated by asserting or retracting rules, thus changing the state of the program. LP updating is appropriate for representing changes and dealing with incomplete information. The specific role of LP updating in our approach is twofold:

1. It updates the program with the preferred explanation to the current observation, thus fixing in the program the initial abduced background context of the counterfactual being evaluated.
2. It facilitates an apposite adjustment to the causal model by hypothetical updates of causal intervention on the program, by affecting defeasible rules in order to retain consistency.

Both roles are sufficiently accomplished by *fluent* (i.e., state-dependent literal) updates, rather than full-blown rule updates. In the first role, explanations are treated as fluents. In the second, reserved predicates are introduced as fluents for the purpose of intervention upon defeasible rules. For the latter role, fluent updates are particularly more appropriate than rule updates (e.g., intervention by retracting rules), because intervention is hypothetical only. Removing away rules from the program would be an overkill, as the rules might be needed to elaborate justifications and introspective debugging. Alternatively, rules can simply be switched off or on in time by means of rule name fluents mechanism, as demonstrated by EVOLP/R.

6.2 Evaluating Counterfactuals via LP Abduction and Updating

The procedure to evaluate counterfactuals in LP essentially takes the three-step process of Pearl's approach as its reference. That is, each step in the LP approach captures the same idea of its corresponding step in Pearl's.

In what follows, counterfactuals are distinguished from semifactuals [4], as the LP procedure for the former is slightly different from the latter. The procedure for semifactuals will be discussed separately at the end of this section.

The key idea of evaluating counterfactuals with respect to an abductive framework, at some current state (discrete time) T, is as follows.

- In step 1, abduction is performed to explain the factual observation.[2] The observation corresponds to the evidence that both the antecedent and the consequence

[2]We assume that people are using counterfactuals to convey truly relevant information rather than to fabricate arbitrary subjunctive conditionals (e.g., "If I had been watching, then I would have seen the cheese on the moon melt during the eclipse"). Otherwise, implicit observations must simply be made explicit observations, to avoid natural language conundrums or ambiguities [10].

literals of the present counterfactual were factually false.[3] There can be multiple explanations available to an observation; choosing a suitable one among them is a pragmatic issue, which can be dealt with preferences or integrity constraints. The explanation fixes the abduced context in which the counterfactual is evaluated ("all other things being equal") by updating the program with the explanation.

- In step 2, defeasible rules are introduced for atoms forming the antecedent of the counterfactual. Given the past event E, that renders its corresponding antecedent literal false, held at factual state $T_E < T$, its causal intervention is realized by a hypothetical update H at state $T_H = T_E + \Delta_H$, such that $T_E < T_H < T_E + 1 \leq T$. That is, a hypothetical update strictly takes place between two factual states, thus $0 < \Delta_H < 1$. In the presence of defeasible rules, this update permits hypothetical modification of the program to consistently comply with the antecedent of the counterfactual.
- Finally, in step 3, the well-founded model of the hypothetical modified program is examined to verify whether the consequence of the counterfactual holds true at state T. One can easily reinstate to the current factual situation by canceling the hypothetical update, e.g., via a new update of H's complement at state $T_F = T_H + \Delta_F$, such that $T_H < T_F < T_E + 1$.

Based on these ideas and analogously to the three-step process of Pearl's, our approach is defined in Procedure 2, abstracting from the above state transition detail (cf. Sect. 7.3.2 for the implementation aspect of this state transition). The following definitions are needed by the procedure.

Definition 6.1 A set of integrity constraint is *satisfied* in $WFM(P)$ iff none is false in $WFM(P)$. That is, the body of an integrity constraint is either false or undefined [21].

We next rephrase Definition 4.15 about abductive solutions and relate them to explanations of observations. As our counterfactual procedure is based on the Well-Founded Semantics, the standard logical consequence relation $P \models F$ used in the definition below presupposes the Well-Founded Model of P in verifying the truth of formula F, i.e., whether F is true in $WFM(P)$.

Definition 6.2 Given an abductive framework $\langle P, \mathcal{AB}, \mathcal{IC} \rangle$ and an observation O, a consistent abductive solution $E \subseteq \mathcal{AB}_{gL}$ is an *explanation* to observation O iff $P \cup E \models O$ and $\mathcal{IC}$ is satisfied in $WFM(P \cup E)$, where all abducibles not appearing in E have been replaced by $\mathbf{u}$, both in P and $\mathcal{IC}$.[4]

Procedure 2 *Let $\langle P, \mathcal{AB}, \mathcal{IC} \rangle$ be an abductive framework, where program P encodes the modeled situation on which counterfactuals are evaluated. Consider a counterfactual:*

"If Pre had been true, then Conc would have been true"

[3]This interpretation is in line with the corresponding English construct, cf. [12], commonly known as *third conditionals*.

[4]This replacement of abducible $A \notin E$ with $\mathbf{u}$ in P and $\mathcal{IC}$ is an alternative but equivalent to adding $A \leftarrow \mathbf{u}$ into $P \cup E$, as foreseen by Definition 4.15.

where Pre and Conc are finite conjunctions of literals.

1. ***Abduction**: Let compl(L) be the negation complement of a literal L. Compute an explanation* $E \subseteq \mathcal{AB}_{gL}$ *to the observation* $O = O_{Pre} \cup O_{Conc} \cup O_{Oth}$*, where:*

   - $O_{Pre} = \{compl(L_i) \mid L_i \text{ is in } Pre\}$;
   - $O_{Conc} = \{compl(L_i) \mid L_i \text{ is in } Conc\}$; *and*
   - O_{Oth} *is other (possibly empty) observations,*

   such that $O_{Oth} \cap (O_{Pre} \cup O_{Conc}) = \emptyset$.

   Update program P with E, obtaining program $P \cup E$.
2. ***Action**: For each literal L in conjunction Pre, introduce a pair of reserved meta-predicates make(B) and make_not(B), where B is the atom in L.*
   These two meta-predicates are introduced for the purpose of establishing causal intervention. That is, they are used to express hypothetical alternative events to be imposed.

   This step comprises two stages:

   a. *Transformation:*
      - *Add rule* $B \leftarrow make(B)$ *to program* $P \cup E$.
      - *Add 'not make_not(B)' to the body of each rule in P whose head is B. If there is no such rule, add rule '*$B \leftarrow not\ make\_not(B)$*' to program* $P \cup E$.

      Let $(P \cup E)_\tau$ *be the resulting transform.*
   b. *Intervention:*
      Update program $(P \cup E)_\tau$ *with literal make(B) or make_not(B), for* $L = B$ *or* $L = not\ B$*, respectively. Assuming that Pre is consistent, make(B) and make_not(B) cannot be imposed at the same time.*
      Let $(P \cup E)_{\tau,\iota}$ *be the program obtained after these hypothetical updates of intervention.*
3. ***Prediction**: Verify whether* $(P \cup E)_{\tau,\iota} \models Conc$ *and* $\mathcal{IC}$ *is satisfied in* $WFM((P \cup E)_{\tau,\iota})$.

This three-step procedure defines *valid* counterfactuals.

Definition 6.3 Let $\langle P, \mathcal{AB}, \mathcal{IC} \rangle$ be an abductive framework, where program P encodes the modeled situation on which counterfactuals are evaluated. The counterfactual

"If *Pre* had been true, then *Conc* would have been true"

is *valid* given observation $O = O_{Pre} \cup O_{Conc} \cup O_{Oth}$ iff O is explained by $E \subseteq \mathcal{AB}_{gL}$, $(P \cup E)_{\tau,\iota} \models Conc$, and $\mathcal{IC}$ is satisfied in $WFM((P \cup E)_{\tau,\iota})$.

Since the Well-Founded Semantics supports top-down query-oriented procedures for finding solutions, checking validity of counterfactuals, i.e., whether their conclusion *Conc* follows (step 3), given the intervened program transform (step 2) with respect to the abduced background context (step 1), in fact amounts to checking in a derivation tree whether query *Conc* holds true while also satisfying $\mathcal{IC}$.

Example 6.1 Recall the example from [4]: *Lightning hits a forest and a devastating forest fire breaks out. The forest was dry after a long hot summer and many acres were destroyed.*

Let us slightly complicate it by having two alternative abductive causes for the forest fire, viz., storm (which implies lightning hitting the ground) or barbecue. Storm is accompanied by strong wind that causes the dry leaves falling onto the ground. Note that dry leaves are important for forest fire in both cases.

This example is expressed by an abductive framework $\langle P, \mathcal{AB}, \mathcal{IC}\rangle$, where P is the program below, $\mathcal{AB} = \{storm/0, barbecue/0\}$, and $\mathcal{IC} = \emptyset$:

$$\begin{aligned} fire &\leftarrow barbecue, dryLeaves \\ fire &\leftarrow barbecue^*, lightning, dryLeaves, leavesOnGround \quad (6.1) \\ lightning &\leftarrow storm \\ leavesOnGround &\leftarrow storm \\ dryLeaves&. \end{aligned}$$

The use of *barbecue** in the body of the rule 6.1 is intended so as to have mutual exclusive explanations. Consider counterfactual:

"If only there had not been lightning, then the forest fire would not have occurred"

where $Pre = not\ lightning$ and $Conc = not\ fire$.

1. ***Abduction***: Besides $O_{Pre} = \{lightning\}$ and $O_{Conc} = \{fire\}$, say that we also observe *leavesOnGround*, i.e., $O_{Oth} = \{leavesOnGround\}$. Given $O = O_{Pre} \cup O_{Conc} \cup O_{Oth}$, there are two possible explanations: $E_1 = \{storm, barbecue^*\}$ and $E_2 = \{storm, barbecue\}$.
Consider a scenario where the minimal explanation E_1 (in the sense of minimal positive literals) is preferred to update P, to obtain $P \cup E_1$. Note, program $P \cup E_1$ corresponds to a state with:

$$WFM(P \cup E_1) = \{dryLeaves, storm, leavesOnGround, lightning, fire, not\ barbecue\}.$$

This updated program reflects the evaluation context of the counterfactual, where all literals of *Pre* and *Conc* were false in the initial factual situation.

2. ***Action***: The transformation results in program $(P \cup E_1)_\tau$:

$$\begin{aligned}
\mathit{fire} &\leftarrow \mathit{barbecue}, \mathit{dryLeaves}.\\
\mathit{fire} &\leftarrow \mathit{barbecue}^*, \mathit{lightning}, \mathit{dryLeaves}, \mathit{leavesOnGround}.\\
\mathit{leavesOnGround} &\leftarrow \mathit{storm}.\\
\mathit{lightning} &\leftarrow \mathit{make}(\mathit{lightning}).\\
\mathit{lightning} &\leftarrow \mathit{storm}, \mathit{not\ make_not}(\mathit{lightning}).\\
\mathit{dryLeaves}&.
\end{aligned}$$

Program $(P \cup E_1)_\tau$ is updated with *make_not(lightning)* as the required intervention, resulting in the program $(P \cup E_1)_{\tau,\iota}$ that corresponds to a new state with:

$$\begin{aligned}
WFM((P \cup E_1)_{\tau,\iota}) = \{&\mathit{dryLeaves}, \mathit{storm}, \mathit{leavesOnGround}, \mathit{make_not}(\mathit{lightning}),\\
&\mathit{not\ make}(\mathit{lightning}), \mathit{not\ barbecue}, \mathit{not\ lightning}, \mathit{not\ fire}\}.
\end{aligned}$$

3. ***Prediction***: We verify that $(P \cup E_1)_{\tau,\iota} \models \mathit{not\ fire}$, and $\mathcal{IC} = \emptyset$ is trivially satisfied in $WFM((P \cup E_1)_{\tau,\iota})$.

We thus conclude that, for this E_1 scenario, the given counterfactual is *valid*.

Example 6.2 In the other explanatory scenario of Example 6.1, where E_2 (instead of E_1) is preferred to update P, the counterfactual is no longer valid, because:

$$\begin{aligned}
WFM((P \cup E_2)_{\tau,\iota}) = \{&\mathit{dryLeaves}, \mathit{storm}, \mathit{leavesOnGround}, \mathit{barbecue},\\
&\mathit{make_not}(\mathit{lightning}), \mathit{not\ make}(\mathit{lightning}), \mathit{not\ lightning}, \mathit{fire}\}
\end{aligned}$$

and thus $(P \cup E_2)_{\tau,\iota} \not\models \mathit{not\ fire}$. Indeed, the forest fire would still have occurred but due to an alternative cause, viz., barbecue.

Skeptical and credulous counterfactual evaluations could ergo be defined, i.e., by evaluating the presented counterfactual for each abduced background context. Given that step 2 can be accomplished by a one-time transformation, such skeptical and credulous counterfactual evaluations require only executing step 3 for each background context fixed in step 1.

Semifactuals Reasoning

Another form related to counterfactuals is *semifactuals*, i.e., one that combines a counterfactual hypothetical antecedent and an unchanged factual consequence [4], with a typical form of statement "Even if ...". Other comparable linguistic constructs also exist, e.g., "No matter if ...", "Though ..., ...still ...", etc. The LP procedure for counterfactuals (Procedure 2) can easily be adapted to evaluating semifactuals. Like in counterfactuals, the antecedent of a semifactual is supposed false in the factual situation. But different from counterfactuals, the consequence of a semifactual should instead be factually ensured *true* (rather than false).

Consider the general semifactual form:

"Even if *Pre* had been true, *Conc* would still have been true".

Its LP evaluation follows Procedure 2 with the only modification on the definition of O_{Conc} in Step 1, i.e., for semifactuals, O_{Conc} is defined as $O_{Conc} = \{L_i \mid L_i \text{ is in } Conc\}$, to warrant its consequence factually true. The validity condition for semifactuals is the same as for counterfactuals, cf. Definition 6.3.

Example 6.3 Recall Example 6.2, where $E_2 = \{storm, barbecue\}$ is preferred. Consider semifactual:

"Even if there had not been lightning, the forest fire would still have occurred"

where $Pre = not\ lightning$ and $Conc = fire$.

This semifactual is valid, because given the same $WFM((P \cup E_2)_{\tau,\iota})$ as in Example 6.2, we now have $(P \cup E_2)_{\tau,\iota} \models Conc$, i.e., $(P \cup E_2)_{\tau,\iota} \models fire$.

6.3 Concluding Remarks

In Pearl's approach, intervention is realized by surface revision, by imposing the desired value to the intervened node and cutting it from its parent nodes. This is also the case in our approach, by means of hypothetical updates affecting defeasible rules that relate to the counterfactual's antecedent. Other subtle ways of intervention may involve deep revision, which can be realized in LP. It is beyond the scope of the book, but amply discussed in [22].

In [22], our procedure is reformulated using different semantics, viz., the weak completion semantics, and some counterfactual properties specific to our LP-based approach are discussed. Indeed, since the idea of each step in the LP approach mirrors the one corresponding in Pearl's, the LP approach therefore immediately compares to Pearl's, its epistemic adequacy and properties relying on those of Pearl's. The satisfaction of counterfactual properties, such as those logic properties of counterfactuals discussed in [15], e.g., various fallacies that distinguish counterfactual conditional from material one, reflexive, modus tollens, disjunction in the antecedent, combination of sentences, etc., is not in the purview of the book, and left for future work.

LP abduction and revision are employed in [7] to evaluate indicative conditionals, but not counterfactual conditionals. LP abduction is employed through a rewrite system to find solutions for an abductive framework; the rewrite system intuitively captures the natural semantics of indicative conditionals. Rule revisions are additionally used to satisfy conditions whose truth-value is unknown and which cannot be explained by abduction. In [6], the rewrite system is extended to evaluate both indicative and subjunctive (or counterfactual) conditionals, with respect to the Weak Completion Semantics. Two notions of *relevance* in the context of the evaluation of conditionals are discussed, viz., weak and strong relevance—both are indirectly inspired by [1, 2]. These two notions of relevance may also be adopted into our counterfactual evaluation procedure; this is left for future work.

The study of counterfactual reasoning in Logic Programming is not new. In [20], counterfactuals are evaluated using contradiction removal semantics of LP. The work is based on Lewis's counterfactuals [15], where a model of a logic program represents a world in Lewis's concept. The semantics defines the most similar worlds by removing contradictions from the associated program, obtaining the so-called maximal non-contradictory submodels of the program. It does not concern itself with LP abduction and updating; both being relevant for our work, which is based on Pearl's concept rather than Lewis's, without the need of a world distance measure.

Probabilistic LP language P-log with the Stable Model Semantics is employed, in [3], to encode Pearl's Probabilistic Causal Model, without involving abduction. It does not directly encode Pearl's three-step process, but focuses on P-log probabilistic approach to compute the probability of a counterfactual query. Our work does not deal with probability, but logic, though it epistemically mirrors Pearl's three-step process, via LP abduction and updating. Our approach is also not based on the stable model semantics, but instead on the Well-Founded Semantics with its relevancy property, which is more appropriate for LP abduction by need as argued earlier.

In [24], Pearl's Probabilistic Causal Model is encoded using a different Probabilistic LP, viz., CP-logic, but without involving abduction either. Whereas P-log has its own *do*-operator to achieve intervention in its probabilistic reasoning, CP-logic achieves it by eliminating rules. Similar to P-log, our approach introduces meta-predicates *make* and *make_not* to accomplish intervention via defeasible rules and fluent updates, without eliminating rules, as CP-logic does.

Our procedure specifically focuses on evaluating counterfactuals in order to determine their validity. It is interesting to explore in future other aspects of counterfactual reasoning; some of them are identified below:

- We consider the so-called *assertive counterfactuals*, where a counterfactual is given as being a valid statement, rather than a statement whose truth validity has to be determined. The causality expressed by such a valid counterfactual may be useful for refining an existing knowledge base. For instance, suppose we have a rule stating that the lamp is on if the switch is on, written as $lamp\_on \leftarrow switch\_on$. Clearly, providing the fact *switch_on*, we have *lamp_on* true. Now consider that the following counterfactual is given as being a valid statement:

  "If the bulb had not functioned properly, then the lamp would not be on"

  There are two ways that this counterfactual may refine the rule about *lamp_on*. First, the causality expressed by this counterfactual can be used to transform the rule into:

$$lamp\_on \leftarrow switch\_on, bulb\_ok.$$
$$bulb\_ok \leftarrow not\ make\_not(bulb\_ok).$$

  So, the lamp will be on if the switch is on—that is still granted—but subject to an update $make\_not(bulb\_ok)$, which captures the condition of the bulb. In the other alternative, an assertive counterfactual is rather directly translated into an update rule, and need not transform existing rules. If we consider an EVOLP-like updating language, the following rule update is enacted:

$$assert(not\ lamp\_on) \leftarrow not\ bulb\_ok$$

and the original rule $lamp\_on \leftarrow switch\_on$ can be kept intact. This rule update affects the query about *lamp_on* thereafter. Like before, the lamp will still be on if the switch is on, but now subject to a superseding *bulb_ok* update.

- We may extend the antecedent of a counterfactual with a rule, instead of just literals. For example, consider the following program (assuming an empty abduction, so as to focus on the issue):

$$\begin{aligned} warm\_blood(M) &\leftarrow mammal(M). \\ mammal(M) &\leftarrow dog(M). \\ mammal(M) &\leftarrow bat(M). \\ dog(d). &\quad bat(b). \end{aligned}$$

Querying `?-` $bat(B), warm\_blood(B)$ assures us that there is a warm blood bat, viz., $B = b$.
Now consider the counterfactual:

"If bats were not mammals they would not have warm blood".

Transforming the above program using our procedure obtains:

$$\begin{aligned} warm\_blood(M) &\leftarrow mammal(M). \\ mammal(M) &\leftarrow make(mammal(M)). \\ mammal(M) &\leftarrow dog(M), not\ make\_not(mammal(M)). \\ mammal(M) &\leftarrow bat(M), not\ make\_not(mammal(M)). \\ dog(d). &\qquad bat(b). \end{aligned}$$

The antecedent of the given counterfactual can be expressed as the rule:

$$make\_not(mammal(B)) \leftarrow bat(B).$$

We can check using our procedure that, given this rule intervention, the above counterfactual is valid: $not\ warm\_blood(b)$ is true in the intervened modified program.

- Finally, we can easily imagine the situation where the antecedent *Pre* of a counterfactual is not given, though the conclusion *Conc* is, and we want to abduce *Pre* in the form of interventions. That is, the task is to abduce *make* and *make_not*, rather than imposing them, while respecting the integrity constraints, such that the counterfactual is valid.
Tabling abductive solutions, such as in TABDUAL, may be relevant in this problem.

Suppose that we already abduced an intervention Pre_1 for a given $Conc_1$, and we now want to find Pre_2 such that the counterfactual "If Pre_1 and Pre_2 had been the case, then $Conc_1$ and $Conc_2$ would have been the case" is valid. In particular, when abduction is performed for a more complex conclusion $Conc_1$ and $Conc_2$, the solution Pre_1, which has already been abduced and tabled, can be reused in the abduction of such a more complex conclusion, leading to the idea that problems of this kind of counterfactual reasoning can be solved in parts or in a modular way.

In summary, this chapter presents a LP technique for evaluating counterfactuals by resorting to a combination of abduction and updating. It corresponds to the three-step procedure of Pearl's structural theory, omitting probability, and focuses on the logical validity of counterfactuals. In future, it is worth exploring possible extensions of this technique to the three counterfactual reasoning aspects discussed above.

References

1. Anderson, A.R., Belnap, N.: Entailment: The Logic of Relevance and Necessity, vol. I. Princeton University Press, New Jersey (1975)
2. Anderson, A., Belnap, N., Dunn, J.: Entailment: The Logic of Relevance and Necessity, vol. II. Princeton University Press, New Jersey (1992)
3. Baral, C., Hunsaker, M.: Using the probabilistic logic programming language P-log for causal and counterfactual reasoning and non-naive conditioning. In: Proceedings of the 20th International Joint Conference on Artificial Intelligence (IJCAI) (2007)
4. Byrne, R.M.J.: The Rational Imagination: How People Create Alternatives to Reality. MIT Press, Cambridge (2007)
5. Collins, J., Hall, N., Paul, L.A. (eds.): Causation and Counterfactuals. MIT Press, Cambridge (2004)
6. Dietz, E.A., Hölldobler, S., Pereira, L.M.: On conditionals. In: Proceedings of Global Conference on Artificial Intelligence (GCAI 2015) (2015)
7. Dietz, E.A., Hölldobler, S., Pereira, L.M.: On indicative conditionals. In: Proceedings of the 1st International Workshop on Semantic Technologies (IWOST), CEUR Workshop Proceedings, vol. 1339 (2015)
8. Epstude, K., Roese, N.J.: The functional theory of counterfactual thinking. Pers. Soc. Psychol. Rev. **12**(2), 168–192 (2008)
9. Ginsberg, M.L.: Counterfactuals. Artif. Intell. **30**(1), 35–79 (1986)
10. Grice, P.: Studies in the Way of Words. Harvard University Press, Cambridge (1991)
11. Halpern, J.Y., Hitchcock, C.: Graded causation and defaults. B. J. Philos. Sci. **66**, 413–457 (2015)
12. Hewings, M.: Advanced Grammar in Use with Answers: A Self-Study Reference and Practice Book for Advanced Learners of English. Cambridge University Press, New York (2013)
13. Hölldobler, S., Ramli, C.D.P.K.: Logic programs under three-valued Łukasiewicz semantics. In: Proceedings of the 25th International Conference on Logic Programming (ICLP), LNCS, vol. 5649, pp. 464–478. Springer, Berlin (2009)
14. Kleiman-Weiner, M., Gerstenberg, T., Levine, S., Tenenbaum, J.B.: Inference of intention and permissibility in moral decision making. In: Proceedings 37th Annual Conference of the Cognitive Science Society (2015)
15. Lewis, D.: Counterfactuals. Harvard University Press, Cambridge (1973)
16. Markman, K.D., Gavanski, I., Sherman, S.J., McMullen, M.N.: The mental simulation of better and worse possible worlds. J. Exp. Soc. Psychol. **29**, 87–109 (1993)

17. McCloy, R., Byrne, R.M.J.: Counterfactual thinking about controllable events. Mem. Cognit. **28**, 1071–1078 (2000)
18. Migliore, S., Curcio, G., Mancini, F., Cappa, S.F.: Counterfactual thinking in moral judgment: an experimental study. Front. Psychol. **5**, 451 (2014)
19. Pearl, J.: Causality: Models. Reasoning and Inference. Cambridge University Press, Cambridge (2009)
20. Pereira, L.M., Aparício, J.N., Alferes, J.J.: Counterfactual reasoning based on revising assumptions. In: Proceedings of the International Symposium on Logic Programming (ILPS 1991), pp. 566–577. MIT Press (1991)
21. Pereira, L.M., Aparício, J.N., Alferes, J.J.: Hypothetical reasoning with well founded semantics. In: Proceedings of the 3rd Scandinavian Conference on Artificial Intelligence. IOS Press (1991)
22. Pereira, L.M., Dietz, E.A., Hölldobler, S.: An abductive counterfactual reasoning approach in logic programming (2015). Available from http://goo.gl/bx0mIZ
23. Roese, N.J.: Counterfactual thinking. Psychol. Bull. **121**(1), 133–148 (1997)
24. Vennekens, J., Bruynooghe, M., Denecker, M.: Embracing events in causal modeling: Interventions and counterfactuals in CP-logic. In: JELIA 2010, LNCS, vol. 6341, pp. 313–325. Springer (2010)

Chapter 7
Logic Programming Systems Affording Morality Experiments

In Chap. 4 we show the appropriateness of LP-based reasoning features for representing diverse issues of moral facets identified in Chap. 3. In this chapter, we discuss how these LP-based reasoning features are synthesized in three different systems: ACORDA (Sect. 7.1), PROBABILISTIC EPA (Sect. 7.2), and QUALM (Sect. 7.3). Whereas the development of QUALM is a contribution of this book, ACORDA [11, 12, 15] and PROBABILISTIC EPA [9, 10, 14] are two existing systems that have been developed earlier, but not with any specific wide principled implementation of morality in mind, as we shall see.

Though these systems share its main feature, viz., abduction, each system concern itself with a particular combination of features. Moreover, their shared feature, abduction, implements different techniques, indicating the progress made in the development of these three systems. The three systems are employed to model, here, for the first time, different issues of considered moral facets, depending on the need of their respective combination of features. Their applications are elaborated in the subsequent Chap. 8.

7.1 ACORDA

ACORDA is a system that implements Prospective Logic Programming [15]. Prospective Logic Programming enables an evolving program to look ahead prospectively into its possible future states and to prefer among them to satisfy goals. This paradigm is particularly beneficial to the agents community, since it can be used to predict an agent's future by employing the methodologies from LP abduction, updating, and preferences, in order to synthesize and maintain abductive hypotheses.

Figure 7.1 [15] shows the architecture of agents that are based on prospective logic. Each prospective logic agent is equipped with a knowledge base and a moral theory as its initial updatable state. The problem of prospection is then of finding abductive solutions to this initial and subsequent states which are both:

L.M. Pereira and A. Saptawijaya, *Programming Machine Ethics*,
Studies in Applied Philosophy, Epistemology and Rational Ethics 26,
DOI 10.1007/978-3-319-29354-7_7

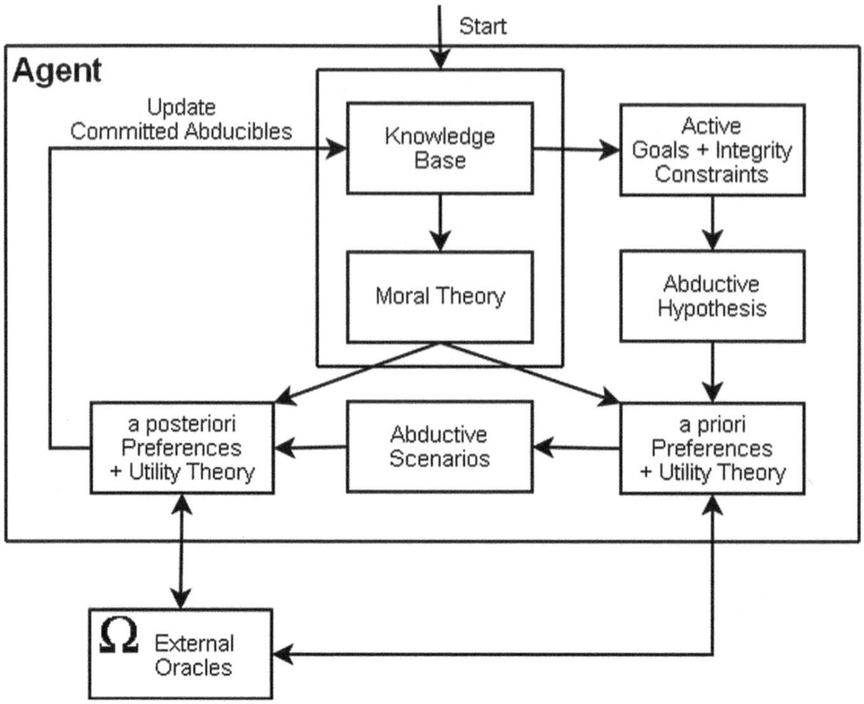

Fig. 7.1 Prospective logic agent architecture (equipped with moral theory)

- *relevant*, i.e., under the agent's current goals; and
- *preferred*, i.e., with respect to preference rules in its knowledge base.

The first step is to select the goals that the agent will attend to during the prospection part of its cycle. Integrity constraints are also considered here to ensure the agent always performs transitions into valid evolution states.

Once the set of active goals for the current state is known, the next step is to find out which are the relevant abductive hypotheses. This step may include the application of a priori preferences, in the form of domain-dependent preference rules, among available abducibles to generate possible abductive scenarios. Forward reasoning can then be applied to the abducibles in those scenarios to obtain relevant consequences, which can then be used to enact a posteriori preferences. These preferences can be enforced by employing utility theory and, in a moral situation, also some available moral theory to be defined. In case additional information is needed to enact preferences, the agent may consult external oracles. This greatly benefits agents in giving them the ability to probe the outside environment, thus providing better informed choices, including the making of experiments. The mechanism to consult oracles is realized by posing questions to external systems, be they other agents, actuators, sensors or ancillary procedures. The agent may use the additional information it acquires to commit to particular abductive solutions, which will be taken into account in a second round of prospection.

ACORDA implements an ad hoc abduction by means of even loops over negation, as described below, on top of an EVOLP meta-interpreter [3]. Its implementation of preferences over abductive scenarios is supported by the results from [8].

ACORDA is implemented in XSB Prolog with its Well-Founded Semantics. Its implementation particularly benefits from the *XSB Answer Set Programming (XASP)* package [6]. The XASP package extends the computation of the well-founded model by providing linkage to the stable model generator (answer set solver) *Smodels* [13] to compute two-valued models from the so-called *residual program* resulting from the top-down query-oriented procedure. The residual program corresponds to a *delay list*, which contains literals whose truth value is undefined in the well-founded model of a program. Such an integration with a stable model generator as provided by the XASP package allows maintaining the relevancy property in finding an answer to a query, by submitting only the relevant residual program to the stable model generator.

We discuss below the main constructs of ACORDA, which are adapted from [15].[1]

7.1.1 Active Goals

In each cycle of its evolution a prospective logic agent attends to a set of active goals. These active goals are triggered by using the observation $on\_observe(O, R, Q)$ construct. This construct generally refers to a relation amongst the observer O, the reporter R, the observation Q, and represents observations that may be reported by the environment to the agent, from one agent to another, or may also be from itself (self-triggered goals).

Being an active goal, triggering $on\_observe/3$ causes the agent to launch the query Q standing for the observations contained inside. In the prospection mechanism, when starting a cycle, the agent collects its active goals by finding all $on\_observe(agent, agent, Q)$ in the program, and subsequently computing abductive solutions, using the mechanism of abduction described below, for the conjunction of Qs obtained from all active goals, while also satisfying integrity constraints. Accordingly, if there is no such active goals in the program, the initial prospection mechanism amounts to satisfying integrity constraints only. In ACORDA, satisfying integrity constraints is realized by invoking the goal $not\ false$, where $false$ is the ACORDA's reserved atom for representing $\bot$ in the Definition 4.12 of integrity constraints.

[1]The initial version of ACORDA is based on [11, 12].

7.1.2 Abduction and A Priori Preferences

In prospective logic, only abducibles that are relevant for the problem in hand are generated. For so doing, the notion of expectation, as described in Sect. 4.3, is employed to express preconditions for enabling the assumption of an abducible, thus constraining a priori relevant abducibles with respect to the agent's actual situation.

An abducible A is only *considered* (and thereby abduced) if there is an expectation for it, and there is no expectation to the contrary. In ACORDA, this is represented as follows:

$$consider(A) \leftarrow expect(A), not\ expect\_not(A), abduce(A).$$

ACORDA implements an ad hoc abduction by means of even loops over negation for every positive abducible A:

$$\begin{array}{ll} abduce(A) & \leftarrow not\ abduce\_not(A). \\ abduce\_not(A) & \leftarrow not\ abduce(A). \end{array}$$

As shown by Example 4.2 in Chap. 4, such representation causes the abducible A undefined in the Well-Founded Model of the program. It creates a relevant residual program in XSB Prolog with respect to the query derived from active goals (and integrity constraints). This relevant residual program can be sent to the stable models generator Smodels through the XASP package, as described above, which will returns abductive stable models that can be analyzed further through a posteriori preferences.

7.1.3 A Posteriori Preferences

Having computed abductive stable models, which correspond with possible scenarios, more favorable ones can be preferred a posteriori. A posteriori preferences are performed to reason about which consequences of abducibles, or other features of the models, are determinant for the final choice, reflecting the desired quality of the model.

One possibility of this a posteriori preferences reasoning is to consider a quantitative evaluation which can be based on different techniques of quantitative decision making. For instance, some measure of utility can be associated to each choice scenario, and the final choice is preferred by maximizing some utility function. Another possibility is to evaluate each choice scenario qualitatively, e.g., by enforcing this scenario to satisfy some properties.

When currently available knowledge of the situation is insufficient to commit to any single preferred abductive solution, additional information can be gathered, e.g., by performing experiments or consulting an oracle, in order to confirm or disconfirm some of the remaining candidates.

7.2 PROBABILISTIC EPA

ACORDA was further developed into the Evolution Prospection Agent (EPA) system [9, 14]. Distinct from ACORDA, EPA considers a different abduction mechanism with a posteriori preferences representation. EPA is subsequently extended with a new feature that allows probabilistic reasoning. The latter is based on the probabilistic logic programming language P-log [5] in a XSB implementation, named P-log(XSB) [10]. We refer to the this extension of EPA as PROBABILISTIC EPA.

The prospection mechanism in EPA is initiated in the same way as in ACORDA, viz., by collecting active goals when a cycle is started, and subsequently finding the abductive solutions of its conjunction, while satisfying integrity constraints. EPA simplifies the representation of an observed active goal G by $on\_observe(G)$, and uses the same reserved atom $false$ for the head of an integrity constraint.

We detail below the features of EPA that mainly distinguish itself from ACORDA.

7.2.1 *Abduction and A Priori Preferences*

In EPA, the reserved predicate $abds(L)$ is used for declaring abducibles (and their corresponding arity) in list L. Like ACORDA, EPA also employs the notion of expectation and considered abducibles for constraining relevant abducibles to the agent's actual situation. Nevertheless, its definition of $consider/1$ is slightly different:

$$consider(A) \leftarrow expect(A),\ not\ expect\_not(A),\ A.$$

where, distinct from ACORDA, the abduction of A is not enacted via an even loop over negation. Instead, the abduction mechanism in EPA is based on the dual program transformation of ABDUAL [4].

Distinct from ACORDA, whose implementation is based on the EVOLP meta-interpreter, the EPA system is implemented on top of the NEGABDUAL meta-interpreter [1]. This meta-interpreter is based on ABDUAL (which is responsible for the abduction mechanism of EPA) but with an additional constructive negation feature. That is, in addition to the abduction mechanism provided by ABDUAL, NEGABDUAL also uses abduction for constructive negation, viz., by making the disunification predicate an abducible. This feature of constructive negation by abduction is beyond the scope of this book, but its discussion is referred to [7]. Given this implementation of EPA, for updating a program with new information, standard Prolog assertion and retraction predicates are used rather than the full EVOLP language [2].

7.2.2 A Posteriori Preferences

EPA introduces a specific syntax for a posteriori preferences:

$$A_i \ll A_j \leftarrow holds\_given(L_i, A_i),\ holds\_given(L_j, A_j). \qquad (7.1)$$

where A_i, A_j are abductive solutions and L_i, L_j are literals representing consequences. Distinct from ACORDA, the a posteriori preferences in EPA does not make use of the XASP package, but resorts instead to abduction itself. In particular, the a posteriori preference rule (7.1) states that A_i is preferred to A_j if L_i and L_j are true as the consequence of abductive solutions A_i and A_j, respectively, without any further abduction being permitted. Optionally, the body of this preference rule may contain Prolog predicate for quantitatively comparing the consequences L_i and L_j.

Other specific evaluations may also figure in the body of the preference rule (7.1). For example, if the evaluation is based on expected utility maximization, then the preference rule can be expressed as:

$$A_i \ll A_j \leftarrow expected\_utility(A_i, U_i),\ expected\_utility(A_j, U_j), U_i > U_j.$$

The rule states that A_i is preferred to A_j if the expected utility value U_i of relevant A_i's consequences is greater than expected utility value U_j of A_j's. Other decision rules, such as maximizing the minimum gain (maximin) or minimizing the maximum possible loss (minimax), are also possible.

7.2.3 Probabilistic Reasoning

PROBABILISTIC EPA extends EPA with a probabilistic reasoning feature based on the probabilistic LP language P-log [5].

The original P-log implementation uses an answer set solver as a tool for computing stable models of its logical part. An alternative implementation of P-log in XSB Prolog, named P-log(XSB) [10], uses the XASP package for interfacing with the answer set solver Smodels. As in ACORDA, this implementation of P-log in XSB Prolog with its underlying Well-Founded Semantics has the advantage of collecting only relevant abducibles for a given query, obtained by need via top-down search, while still benefiting from the computation of stable models through the XASP package. Moreover, the tabling mechanism in XSB Prolog may significantly improve the performance of P-log(XSB) compared to the original P-log implementation, as shown by the evaluation results in [10].

PROBABILISTIC EPA results from the integration of P-log(XSB) into EPA. We next summarize the components of P-log and their syntax in PROBABILISTIC EPA.

A P-log program Π consists of a sorted signature, declarations, a regular part, a set of random selection rules, a probabilistic information part, and a set of observations and actions.

Sorted Signature

The sorted signature Σ of Π contains a set of constant symbols and term-building function symbols, which are used to form terms in the usual way. Additionally, the signature contains a collection of special function symbols called *attributes*. Attribute terms are expressions of the form $a(\bar{t})$, where a is an attribute and $\bar{t}$ is a vector of terms of the sorts required by a.

Declaration

The declaration part of a P-log program can be defined as a collection of sorts and sort declarations of attributes. A sort c can be defined by listing all the elements $c = \{x_1, \ldots, x_m\}$ or by specifying the range of values $c = \{L \ldots U\}$, where L and U are the integer lower bound and upper bound of the sort c. Attribute a with domain $c_1 \times \cdots \times c_n$ and range c_0 is represented as follows:

$$a : c_1 \times \cdots \times c_n \text{ -> } c_0.$$

If attribute a has no domain parameter, we simply write $a : c_0$. The range of attribute a is denoted by range(a).

Regular Part

This part of a P-log program consists of a collection of rules, facts and integrity constraints, formed using literals of Σ.

Random Selection Rule

This is a rule for attribute a, which has the form:

$$random(RandomName, a(\bar{t}), DynamicRange) \;:\!-\; Body.$$

This means that the attribute instance $a(\bar{t})$ is random if the conditions in $Body$ are satisfied. The parameter $DynamicRange$ allows restricting the default range for random attributes. The parameter $RandomName$ is a syntactic mechanism used to link random attributes to the corresponding probabilities. A constant $full$ can be used in $DynamicRange$ to signal that the dynamic range is equal to $range(a)$.

Probabilistic Information

Information about probabilities of random attribute instances $a(\bar{t})$ taking a particular value y is given by probability atoms (called pa-atoms), which can be defined by the following pa-rule:

$$pa(RandomName, a(\bar{t}, y), d\_(A, B)) \;:\!-\; Body.$$

This rule expresses that if the *Body* were true, and the value of $a(\bar{t})$ were selected by a rule named *RandomName*, then *Body* would cause $a(\bar{t}) = y$ with probability $\frac{A}{B}$. Note that the probability of an atom $a(\bar{t}, y)$ will be directly assigned if the corresponding *pa*-atom is the head of some *pa*-rule, whose body is true. To define probabilities of the remaining atoms we assume that, by default, all values of a given attribute which are not assigned a probability are equally likely.

Observations and Actions

These are, respectively, statements of the forms $obs(l)$ and $do(l)$, where l is a literal. Observations $obs(a(\bar{t}, y))$ are used to record the outcome y of a random event $a(\bar{t})$. Statement $do(a(\bar{t}, y))$ indicates $a(\bar{t}) = y$ is enforced as the result of a deliberate action.

In a PROBABILISTIC EPA program, a P-log program is embedded by putting it between reserved keywords, beginPlog and endPlog. In PROBABILISTIC EPA, probabilistic information can be obtained using the P-log(XSB) reserved predicate $pr(Q, P)$ [10], which computes the probability P of a given query Q. It can be embedded just like a usual Prolog predicate, in any constructs of PROBABILISTIC EPA programs, including active goals, preferences, and integrity constraints. Moreover, since P-log(XSB) allows to code Prolog probabilistic meta-predicates, i.e., Prolog predicates that depend on $pr/2$ predicates, we can directly use this probabilistic meta-information in PROBABILISTIC EPA programs.

7.3 QUALM

In Chap. 5, we propose two original approaches that enjoy the benefit of tabling for different reasoning features, separately, viz., for abduction (TABDUAL) and for updating (EVOLP/R). The integration of these two approaches is indispensable for our purpose of representing some moral facets, as argued in Chap. 4. This section discusses a unified approach, realized in a system prototype QUALM, to seamlessly integrate TABDUAL and EVOLP/R by joint tabling of abduction and updating, in order to keep the benefit of tabling in each individual approach. Additionally, QUALM implements our LP-based counterfactual evaluation procedure that requires the interplay between abduction and updating, as described in Chap. 6.

7.3.1 Joint Tabling of Abduction and Updating

Our initial attempt, in [17], depends on the notion of expectation, which is also employed in ACORDA and EPA, to express preconditions for enabling the assumption of an abducible. This concept consequently requires every rule with abducibles in its body to be preprocessed, by substituting abducible A with $consider(A)$.

While this notion of expectation is useful in constraining abducibles to those relevant for the problem at hand, its application in [17] limits the benefit of tabling abductive solutions. That is, instead of associating tabling abductive solutions to an atom, say q (via a tabled predicate q_{ab}), a single and generic tabled predicate $consider_{ab}(A)$ is introduced for tabling just this one single given abducible A under consideration. Even though $expect(A)$ and $expect\_not(A)$ may have rules stating conditions for expecting A or otherwise, it is unnatural that such rules would have other abducibles as a condition for abducing A, as $consider/1$ concerns conditions for *one* abducible only, according to its intended semantics and use. That is, a $consider/1$ call must not depend on another $consider/1$, and so will never have an abductive context. Such a generic tabling by $consider_{ab}$ is superfluous and not intended by the concept of tabling abductive solutions introduced in TABDUAL. In this section we remedy the joint tabling approach of [17] in order to uphold the idea and the benefit of tabling abductive solutions in contextual abduction.

The joint tabling approach in combining the techniques from TABDUAL and EVOLP/R naturally depends on two pieces of information carried from each of these techniques, viz. abductive contexts and the timestamp (i.e., holds-time) indicating when a fluent holds true, respectively. They keep the same purpose in the integration as in their respective individual approach. Example 7.1 shows how both entries figure in a remedied rule transform.

Example 7.1 Recall rule $s \leftarrow q, b$ in Example 5.1 of Chap. 5. Its rule transform below illustrates the joint tabling approach by combining the abductive contexts and the holds-time information:

$$s_{ab}(E_3, H) \leftarrow \#r(s, [q, b], [\,], E_1, H_r),\ b(E_1, E_2, H_b),\ q(E_2, E_3, H_q),$$
$$latest([(\#r(s, [q, b]), H_r), (b, H_b), (q, H_q)], H).$$

where $s_{ab}/2$ is now an *incremental* tabled predicate, which later can be reused in the definition of $s/3$ (see rule (7.2) below).

There are three important points in this transformation to clarify.

- Unlike the approach in [17], the abducible b in the body is not required to be preprocessed into $consider(b)$. Furthermore, the abducible b is now called explicitly, rather than immediately placed as the input abductive context of q as in the TABDUAL transformation (cf. rule (5.3) in Chap. 5).
  This explicit call is intended to anticipate possible updates on this abducible. Such abducible updates may take place when one wants to commit to a preferred explanation and fix it in the program as a fact. For example, this is the case when an abduced background context of a counterfactual needs to be fixed, as shown in our counterfactual evaluation procedure in Chap. 6. Having an abducible as an explicit goal in the body thus facilitates bottom-up propagation of its updates, which is induced by incremental tabling of fluents, due to the dependency of the tabled predicate $s_{ab}/2$ in the head on the abducible goal in the body ($b/3$). This explicit

call of abducible is resolved by the transformation of abducibles, similar to that of TABDUAL, cf. Definition 5.3 in Chap. 5, e.g., the transform rule for abducible b is as follows:

$$b(I, O, H) \leftarrow insert\_abducible(b, I, O, H)$$

- Like in EVOLP/R, the rule name fluent $\#r(s, [q, b])$ is assigned to uniquely identify the rule $s \leftarrow q, b$, which now also has additional parameters of input and output abductive contexts besides its usual timestamp parameter H_r. Like in TABDUAL, E_3 is an abductive solution tabled via predicate s_{ab}, and obtained from relaying the ongoing abductive solution in context E_1 from the goal $\#r(s, [q, b])$ to the goal q in the body, given the empty input abductive context $[\ ]$ of $\#r(s, [q, b])$. This empty input abductive context is due to the treatment of the abducible b, which now becomes an explicit goal in the body rather than appear in an input abductive context (as explained above).
  The reserved predicate $latest/2$ is borrowed from its EVOLP/R part, which determines the holds-time H of s_{ab} from the three goals in its body that latest holds. Note that if H_b is still a variable then it is set to the latest H.
- By prioritizing abducible goals to occur before any non-abducible goals in the body, the benefit of TABDUAL, viz., of reusing tabled abductive solutions from one context to another, can still be obtained. In Example 7.1, calling the abducible b, before q, with the empty input abductive context, provides q with an actual input abductive context $E_2 = [b]$. It thus achieves the same effect as simply having $[b]$ as the input context of q (cf. rule (5.3) in Chap. 5). Note that the rule name fluent $\#r(s, [q, b])$ is a fact, which transforms into $\#r(s, [q, b], I, I, 1)$, and therefore the empty context is just relayed intact to E_1.

The tabled solution in s_{ab} can be reused via the definition s below, which is similar to rule (5.4) in Chap. 5, except for the addition of timestamp information T:

$$s(I, O, T) \leftarrow s_{ab}(E, T), produce\_context(O, I, E). \tag{7.2}$$

where $produce\_context/3$ is defined as in TABDUAL.

Finally, the different purposes of the dual program transformation, employed both in TABDUAL and EVOLP/R, are consolidated within this unified approach. The abductive contexts and the timestamp information also jointly figure in the parameters of dual predicates, as shown in Example 7.2 below.

Example 7.2 Recall Example 5.4 in Chap. 5. Considering $p/0$ as a fluent, the dual program transformation of the unified approach results in rules below. Note that each rule of $p/0$ is assigned a unique rule name.

$$not\_p(T_0, T_2, H_p) \leftarrow p^{*1}(T_0, T_1, D_{p_1}, H_{p_1}),\ p^{*2}(T_1, T_2, D_{p_2}, H_{p_2}),$$
$$latest([(D_{p_1}, H_{p_1}), (D_{p_2}, H_{p_2})], H_p) \tag{7.3}$$

$$p^{*1}(I, O, not\_\#r(p, [a]), H_{p_{11}}) \leftarrow not\_\#r(p, [a], I, O, H_{p_{11}}).$$

$$p^{*1}(I, O, a^*, H_{p_{12}}) \leftarrow a^*(I, O, H_{p_{12}}).$$

$$p^{*2}(I, O, not\_\#r(p, [q, not\ r]), H_{p_{21}}) \leftarrow not\_\#r(p, [q, not\ r], I, O, H_{p_{21}}).$$
$$p^{*2}(I, O, not\_q, H_{p_{22}}) \leftarrow not\_q(I, O, H_{p_{22}}).$$
$$p^{*2}(I, O, r, H_{p_{23}}) \leftarrow r(I, O, H_{p_{23}}).$$

In each second layer dual rule p^{*i}, the chosen dualized negated literal needs to be passed to the first layer dual rule, in its parameter D_{p_i}. Like in EVOLP/R, this information is needed by $latest/2$ to ensure that none of negated dualized literals in the body were subsequently supervened by their complements at some time before H_p. Indeed, passing this information achieves the same effect as if the dual rules are represented in a flattened form, i.e., if the goals p^{*1} and p^{*2} in the body of not_p are substituted by their chosen dualized negated literals.

7.3.2 Evaluating Counterfactuals

For evaluating counterfactuals, QUALM provides the construct $intervened(L)$ to declare all predicates, in list L, that are subject to intervention. For example, we have $intervened([lightning/0])$ for the only counterfactual in Example 6.1 of Chap. 6. This declaration is useful to determine which rules to transform, according to the step 2 of our counterfactuals evaluation procedure (see Procedure 2 in Chap. 6). The transformation stage in step 2 (but not the intervention) can therefore be performed in advance, as a one-time transformation.

In QUALM, the state transition of the program, as a consequence of program updating (by asserting or retracting fluents, in our case), is facilitated by timestamps that are internally managed. Following the convention of EVOLP/R, the program is initially inserted at state (timestamp) $T = 1$, which subsequently progresses to $T = 2$ as the current state.

Starting in step 1 of Procedure 2, a top-level query $query(Query, In, Out)$ is invoked for finding the explanation, in the output abductive context Out, of the observation $Query$, given the input abductive context In.[2] For example:

$$?\text{-}\ \ query((lightning, fire, leavesOnGround), [\ ], E)$$

provides an explanation E (for an empty input abductive context) to the observation $O = \{lightning, fire, leavesOnGround\}$ of Example 6.1. In order to fix $E = \{storm, barbecue^*\}$ as the abduced background context in evaluating counterfactual at the present state $T = 2$, both fluents $storm$ and $barbecue^*$, that were true at the factual state $T_E = 1$, are asserted. QUALM provides a reserved predicate $updates(L)$ to record pending incremental assertion of fluents (and their assertion timestamps)

[2]Predicate $query/3$ does not explicitly specify a parameter for query-time. In this case, the query-time always refers to the current timestamp. Alternatively, QUALM also provides $query(Query, In, Out, QTime)$, which allows specifying a particular query-time $QTime$.

in the list L. As in EVOLP/R, only those pending assertions whose timestamps up to some query-time will become actual, and this is triggered by a top-level query with its specific query-time.

Providing the transformation stage in step 2 has been performed in advance, the causal intervention "there had not been lightning" is enacted by the hypothetical update of fluent $make\_not(lightning)$, via $updates([make\_not(lightning)])$. As described in Sect. 6.2, this update strictly takes place between two consecutive factual states; in this case between $T_E = 1$ and the current state $T = 2$. QUALM internally assigns a fraction of timestamp, say 0.01, just after T_E, viz., the hypothetical update $make\_not(lightning)$ is imposed at state $T_H = 1.01$. It thus simulates an intervention via an update in the past, while keeping the current state at $T = 2$.

After this update, the validity of the present counterfactual (at $T = 2$) can be checked by testing its conclusion (step 3 of the procedure). For example, the top-level query ?- $query(fire, [\ ], E)$ is launched to ascertain whether forest fire would have occurred after the hypothetical update. QUALM answers 'no', which verifies the counterfactual's validity that the forest fire would not have occurred.

Finally, to reinstate the current factual situation from a counterfactual mode, the previous hypothetical update can be canceled by updating the program with its fluent complement. Continuing the above example, $updates([not\ make\_not(lightning)])$ is given, and QUALM will internally assign a fraction of time after T_H for the timestamp T_F of this update, e.g., at $T_F = T_H + 0.01 = 1.02$. It thus supervenes the hypothetical update $make\_not(lightning)$ that was enacted at $T_H = 1.01$, and consequently, the intervention is no longer imposed on the program.

By using subsequent fractional timestamps to the first counterfactual, other counterfactuals may be queried assuming the hypothetical context of the previous ones, until the whole hypothetical situation is supervened by the above mechanism.

7.4 Concluding Remarks

As we show in the subsequent chapter, the different combinations of features in these three systems allows us to focus on particular morality issues to model. Note that while the three systems discussed here afford morality experiments, their applications clearly are not specific to morality.

The use of the XASP package in ACORDA and PROBABILISTIC EPA is important, as it shows that stable models are computed with respect to a residual program, the latter being itself obtained via a top-down query-oriented procedure (which maintains the relevancy property in finding an answer to a query). While PROBABILISTIC EPA is based on P-log for its probabilistic reasoning, it would be interesting in future to consider alternatives. In particular, since PROBABILISTIC EPA is implemented in XSB Prolog, the *Probabilistic Inference with Tabling and Answer subsumption (PITA)* [16] package in XSB Prolog may be a good alternative candidate. In fact, PITA does not only support probabilistic logic programs, but may also be suitable for possibilistic logic programs. Such available options may benefit PROBABILISTIC EPA for its more general purpose applications.

QUALM is an ongoing work and continuously being improved. It is tempting to add features existing in the other two systems, e.g., preferences or probabilistic reasoning, into QUALM, in order to have a fully integrated system. But this should be prudently considered, as it will obviously increase the complexity of the system. We touch upon this general issue in Chap. 11.

References

1. Alferes, J.J., Pereira, L.M.: NEGABDUAL meta-interpreter (2007). http://centria.di.fct.unl.pt/~lmp/software/contrNeg.rar
2. Alferes, J.J., Brogi, A., Leite, J.A., Pereira, L.M.: Evolving logic programs. In: Proceedings of the European Conference on Artificial Intelligence (JELIA 2002). LNCS, pp. 50–61. Springer, New York (2002)
3. Alferes, J.J., Brogi, A., Leite, J.A., Pereira, L.M.: EVOLP meta-interpreter (2002). http://centria.di.fct.unl.pt/~jja/updates
4. Alferes, J.J., Pereira, L.M., Swift, T.: Abduction in well-founded semantics and generalized stable models via tabled dual programs. Theory Pract. Log. Program. **4**(4), 383–428 (2004)
5. Baral, C., Gelfond, M., Rushton, N.: Probabilistic reasoning with answer sets. Theory Pract. Log. Program. **9**(1), 57–144 (2009)
6. Castro, L., Swift, T., Warren, D.S.: XASP: answer set programming with XSB and Smodels. The XSB System Version 3.6 Manual, vol. 2, Libraries, Interfaces, and Packages (2015)
7. Ceruelo, V.P.: Negative non-ground queries in well founded semantics. Master's thesis, Universidade Nova de Lisboa (2009)
8. Dell'Acqua, P., Pereira, L.M.: Preferential theory revision. J. Appl. Log. **5**(4), 586–601 (2007)
9. Han, T.A.: Evolution prospection with intention recognition via computational logic. Master's thesis, Technische Universität Dresden and Universidade Nova de Lisboa (2009)
10. Han, T.A., Ramli, C.D.K., Damásio, C.V.: An implementation of extended P-log using XASP. In: Proceedings of the 24th International Conference on Logic Programming (ICLP). LNCS, vol. 5366. Springer, New York (2008)
11. Lopes, G.: A computational approach to introspective consciousness in logic programming: ACORDA. Master's thesis, Universidade Nova de Lisboa (2006)
12. Lopes, G., Pereira, L.M.: Prospective programming with ACORDA. In: Empirically Successful Computerized Reasoning (ESCoR 2006) Workshop, IJCAR (2006)
13. Niemelä, I., Simons, P.: Smodels: an implementation of the stable model and well-founded semantics for normal LP. In: Proceedings of the 4th International Conference on Logic Programming and Nonmonotonic Reasoning (LPNMR). LNAI, vol. 1265 (1997)
14. Pereira, L.M., Han, T.A.: Evolution prospection. In: Proceedings of the 1st KES International Symposium on Intelligent Decision Technologies (IDT), vol. 199, pp. 139–150 (2009)
15. Pereira, L.M., Lopes, G.: Prospective logic agents. Int. J. Reason.-Based Intell. Syst. **1**(3/4), 200–208 (2009)
16. Riguzzi, F., Swift, T.: The PITA system: tabling and answer subsumption for reasoning under uncertainty. Theory Pract. Log. Program. **11**(4–5), 433–449 (2011)
17. Saptawijaya, A., Pereira, L.M.: Joint tabling of logic program abductions and updates (Technical communication of ICLP 2014). Theory Pract. Log. Program. **14**(4–5) (2014). Online Supplement http://arxiv.org/abs/1405.2058

Chapter 8
Modeling Morality Using Logic Programming

This chapter aims at realizing our conception about representing diverse moral facets in Logic Programming, by modeling several issues pertaining to those moral facets, using the three systems discussed in Chap. 7. The applicability of these systems corresponds with their relevance to the moral issues being modeled.

In Sect. 8.1, ACORDA is employed to model moral permissibility, emphasizing the use of integrity constraints in abduction and preferences over abductive scenarios, where several cases of the classic trolley problem are modeled. Then, moral reasoning concerning uncertain actions is modeled, in Sect. 8.2, by means of PROBABILISTIC EPA. Finally, we demonstrate the use of QUALM for modeling the issue of moral updating and counterfactual moral reasoning, in Sect. 8.3.

8.1 Moral Reasoning with ACORDA

In Chap. 3, several cases built from the classic trolley problem [4] are presented. The cases concern themselves with the question of moral permissibility in a type of dilemma that involves harm. They are apparently similar in the dilemma they introduce (five people vs. one person being killed), yet the nuances in their specific scenarios and applied moral principles may influence moral permissibility judgments.

We model each case of the trolley problem in ACORDA separately, and demonstrate how appropriate moral decisions are delivered through abductive reasoning. By appropriate moral decisions we mean the ones that conform with those the majority of people make, based on empirical results in the literature. We particularly refer to [8, 9, 17], where experiments are conducted to assess moral judgments of subjects from demographically diverse populations (across gender, ethnicity, religion, age, exposure to moral coursework, etc.) on the following various trolley problem cases. The experiments are specifically intended to support their framework of human moral cognition, known as universal moral grammar [9, 17], analogously to Chomsky's universal grammar in language. It provides universal moral principles, albeit

L.M. Pereira and A. Saptawijaya, *Programming Machine Ethics*,
Studies in Applied Philosophy, Epistemology and Rational Ethics 26,
DOI 10.1007/978-3-319-29354-7_8

being culturally adjustable, that enable an individual to evaluate what actions are permissible, obligatory, or forbidden.

We first describe various cases of the trolley problem. All cases present a moral dilemma that inquires whether it is permissible to harm an individual for the purpose of saving others. The initial circumstances are the same [9]:

There is a trolley and its conductor has fainted. The trolley is headed toward five people walking on the track. The banks of the track are so steep that they will not be able to get off the track in time.

Given the above initial circumstance, we consider below six cases of moral dilemmas [17]. These six cases are described below and visually depicted in Fig. 8.1.

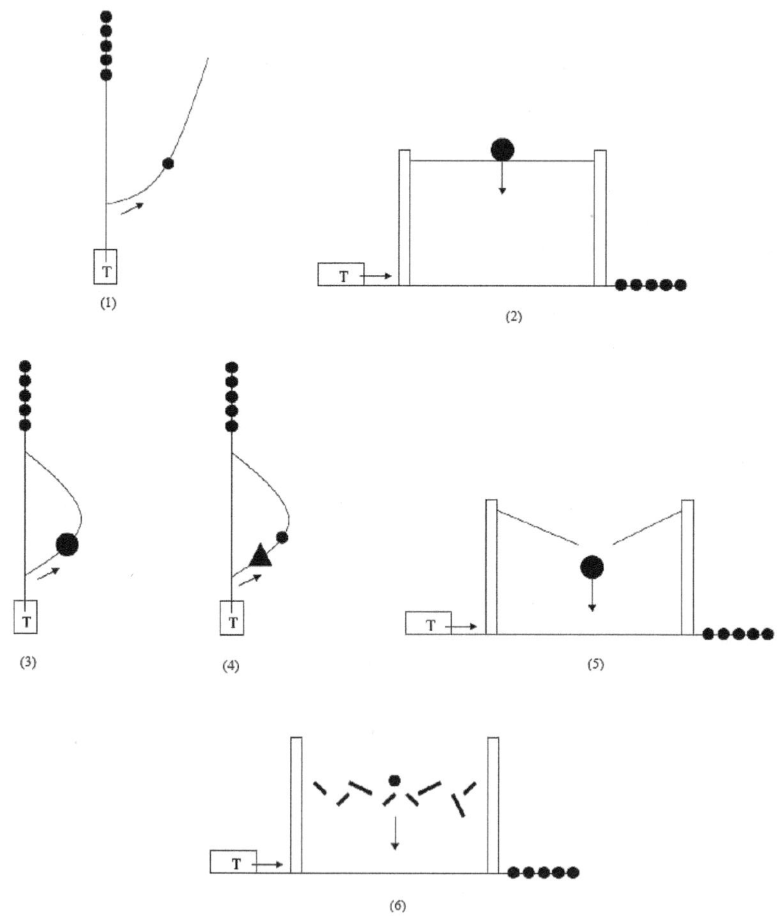

Fig. 8.1 The six trolley cases: (*1*) Bystander, (*2*) Footbridge, (*3*) Loop, (*4*) Man-in-front, (*5*) Drop Man, (*6*) Collapse Bridge

1. **Bystander**. Hank is standing next to a switch, which he can throw, that will turn the trolley onto a parallel side track, thereby preventing it from killing the five people. However, there is a man standing on the side track with his back turned. Hank can throw the switch, killing him; or he can refrain from doing this, letting the five die. Is it morally permissible for Hank to throw the switch?
2. **Footbridge**. Ian is on the footbridge over the trolley track. He is next to a heavy man, which he can shove onto the track in the path of the trolley to stop it, thereby preventing it from killing the five people. Ian can shove the man onto the track, resulting in death; or he can refrain from doing this, letting the five die. Is it morally permissible for Ian to shove the man?
3. **Loop**. Ned is standing next to a switch, which he can throw, that will temporarily turn the trolley onto a loop side track. There is a heavy object on the side track. If the trolley hits the object, the object will slow the train down, giving the five people time to escape. The heavy object is a man. Ned can throw the switch, preventing the trolley from killing the five people, but killing the man. Or he can refrain from doing this, letting the five die. Is it morally permissible for Ned to throw the switch?
4. **Man-in-front**. Oscar is standing next to a switch, which he can throw, that will temporarily turn the trolley onto a looping side track. There is a heavy object on the side track. If the trolley hits the object, the object will slow the train down, giving the five people time to escape. There is a man standing on the side track in front of the heavy object. Oscar can throw the switch, preventing the trolley from killing the five people, but killing the man. Or he can refrain from doing this, letting the five die. Is it morally permissible for Oscar to throw the switch?
5. **Drop Man**. Victor is standing next to a switch, which he can throw, that will drop a heavy object into the path of the trolley, thereby stopping the trolley and preventing it from killing the five people. The heavy object is a man, who is standing on a footbridge overlooking the track. Victor can throw the switch, killing him; or he can refrain from doing this, letting the five die. Is it morally permissible for Victor to throw the switch?
6. **Collapse Bridge**. Walter is standing next to a switch, which he can throw, that will collapse a footbridge overlooking the tracks into the path of the trolley, thereby stopping the train and preventing it from killing the five people. There is a man standing on the footbridge. Walter can throw the switch, killing him; or he can refrain from doing this, letting the five die. Is it morally permissible for Walter to throw the switch?

Interestingly, despite the same dilemma, viz., to save five albeit killing one, subjects come to different judgments on whether the action to reach the goal is permissible or impermissible, as summarized in Table 8.1 (by taking the majority from the result of each case as presented in [17]). Even though the subjects are unable to explain the moral principles in their attempts at justification, their moral judgments are consistent with the Doctrine of Double Effect [16]. This moral principle addresses the permissibility of an action that causes a harm by distinguishing whether this harm is a mere *side-effect* of bringing about a good result (in which case it is permissible),

Table 8.1 Summary of moral judgments for the six trolley problem cases

Trolley problem case	Moral judgment
1. Bystander	Permissible
2. Footbridge	Impermissible
3. Loop	Impermissible
4. Man-in-front	Permissible
5. Drop Man	Impermissible
6. Collapse Bridge	Permissible

or rather an *intended means* to bringing about the same good end (in which case it is impermissible).

The Doctrine of Double Effect is modeled subsequently via integrity constraints and a posteriori preferences to capture the deontological and the utilitarian flavours of moral judgments, respectively, in these cases. Possible decisions are modeled as abducibles. Moral decisions are therefore made by first satisfying integrity constraints (to rule out impermissible actions), then computing abductive stable models from the resulting abductive scenarios, and finally preferring amongst them (by means of LP rules) on the basis of their consequences of abducibles in the models.

In addition to the Doctrine of Double Effect, the Doctrine of Triple Effect [11] is also considered, which extends our LP representation of the Loop Case—the case affected by this doctrine—and contrasts moral permissibility of this specific case with respect to the Doctrine of Double Effect.

Next, we detail the modeling of the above six cases in ACORDA. In each case of the trolley problem, there are always two possible decisions to make. One of these is the same for all cases, i.e. letting the five people die by merely watching the train go straight. The other decision depends on the cases, e.g. throwing the switch, shoving a heavy man, or the combination of them. In order to assess how flexible is our model of the moral rule, we additionally model other variants for the Footbridge and Loop cases.

Modeling the Bystander Case

This case is modeled as follows:

$$
\begin{array}{lll}
\mathit{side\_track}. & \mathit{on\_side}(\mathit{john}). & \mathit{human}(\mathit{john}). \\
\mathit{expect}(\mathit{watch}). & & \\
\mathit{train\_straight} & \leftarrow \mathit{consider}(\mathit{watch}). & \\
\mathit{end}(\mathit{die}(5)) & \leftarrow \mathit{train\_straight}. & \\
\mathit{observed\_end} & \leftarrow \mathit{end}(X). & \\
\mathit{expect}(\mathit{throw\_switch}) & \leftarrow \mathit{side\_track}. & \\
\mathit{turn\_side} & \leftarrow \mathit{consider}(\mathit{throw\_switch}). & \\
\mathit{kill}(1) & \leftarrow \mathit{human}(X), \mathit{on\_side}(X), \mathit{turn\_side}. & \\
\mathit{end}(\mathit{save\_men}, \mathit{ni\_kill}(N)) & \leftarrow \mathit{turn\_side}, \mathit{kill}(N). & \\
\mathit{observed\_end} & \leftarrow \mathit{end}(X, Y). &
\end{array}
$$

The first three facts describe that there is a side track and a man (here, named *john*) standing on that track. The fact $expect(watch)$ and rule $expect(throw\_switch) \leftarrow side\_track$ indicate that watching and throwing the switch, respectively, are two available abducibles, representing possible decisions Hank has. In this case, the action of throwing switch is only expected as an abducible, if the side track exists. The other clauses represent the chain of actions and consequences for these two abducibles.

The predicate $end(die(5))$ represents the final consequence of abducing *watch*, that five people die. On the other hand, the predicate $end(save\_men, ni\_kill(N))$ represents the final consequence of abducing *throw_switch*: it will save the five without intentionally killing someone. The way of representing these two consequences is chosen differently, because these two abducibles are of different kind. Merely watching the trolley go straight is an omission of action that just has negative consequence, whereas throwing the switch is an action that is performed to achieve a goal and additionally has negative consequence. Since abducibles in other cases of the trolley problem also share this property, this way of representing will be used in other cases too. Finally, *observed_end* is used for encapsulating both representations into one, to serve as a goal that will later be triggered by an integrity constraint.

Modeling the Footbridge Case

The program below models the action of shoving an object as an abducible, together with its chain of consequences. The part for the decision of merely watching is the same as in the case of Bystander.

$$
\begin{array}{lll}
stand\_near(john). & human(john). & heavy(john).\\
expect(shove(X)) & \leftarrow stand\_near(X). & \\
on\_track(X) & \leftarrow consider(shove(X)). & \\
stop\_train(X) & \leftarrow on\_track(X), heavy(X). & \\
kill(1) & \leftarrow human(X), on\_track(X). & \\
kill(0) & \leftarrow inanimate\_object(X), on\_track(X). & \\
end(save\_men, ni\_kill(N)) & \leftarrow inanimate\_object(X), stop\_train(X), kill(N). & \\
end(save\_men, i\_kill(N)) & \leftarrow human(X), stop\_train(X), kill(N). & \\
observed\_end & \leftarrow end(X, Y). &
\end{array}
$$

The fact of a heavy man (here, also named *john*) on the footbridge standing near to the agent is modeled similarly as in the Bystander case. This case can be made more interesting by additionally having another inanimate heavy object (e.g., rock) on the footbridge near to Ian, which corresponds to the first variant of the Footbridge case:

$$stand\_near(rock). \quad inanimate\_object(rock). \quad heavy(rock).$$

Alternatively, in the second variant, an even loop over default negation can be used, to represent that either a man or an inanimate object be on the footbridge next to Ian:

$$stand\_near(john) \leftarrow not\ stand\_near(rock).$$
$$stand\_near(rock) \leftarrow not\ stand\_near(john).$$

Note that the action of shoving an object is only possible if there is an object near Ian to shove, hence the rule $expect(shove(X)) \leftarrow stand\_near(X)$. We also have two clauses that describe two possible final consequences. The rule with the head $end(save\_men, ni\_kill(N))$ deals with the consequence of reaching the goal, viz., saving the five, but not intentionally killing someone (in particular, without killing anyone in this case). On the other hand, the rule with the head $end(save\_men, i\_kill(N))$ expresses the consequence of reaching the goal but involving an intentional killing.

Modeling the Loop Case

We consider three variants for the loop track case. In the first variant, instead of having only one looping side track as in the original scenario, we consider two looping side tracks, viz., the left and the right loop side tracks. John, a heavy man, is standing on the left side track, whereas on the right side track there is an inanimate heavy object, e.g., rock. These facts can be represented as follows:

$$\begin{array}{lll} side\_track(left). & side\_track(right). & \\ on(john, left). & human(john). & heavy(john). \\ on(rock, right). & inanimate\_object(rock). & heavy(rock). \end{array}$$

The switch can be thrown to either one of the two looping side tracks, represented with the abducible predicate $throw\_switch/1$. The expectation of this abducible together with the chain of consequences are shown below:

$$\begin{array}{ll} expect(throw\_switch(Z)) & \leftarrow side\_track(Z). \\ turn\_side(Z) & \leftarrow consider(throw\_switch(Z)). \\ slowdown\_train(X) & \leftarrow turn\_side(Z), on(X, Z), heavy(X). \\ kill(1) & \leftarrow turn\_side(Z), on(X, Z), human(X). \\ kill(0) & \leftarrow turn\_side(Z), on(X, Z), inanimate\_object(X). \\ end(save\_men, ni\_kill(N)) & \leftarrow inanimate\_object(X), slowdown\_train(X), kill(N). \\ end(save\_men, i\_kill(N)) & \leftarrow human(X), slowdown\_train(X), kill(N). \\ observed\_end & \leftarrow end(X, Y). \end{array}$$

In the second variant, we consider a single looping side track with either a man or an inanimate object on it. As in the footbridge case, rather than having two separate programs to model these distinct facts, we can model it instead using an even loop over default negation to capture alternatives between objects:

$$side\_track(john) \leftarrow not\ side\_track(rock).$$
$$side\_track(rock) \leftarrow not\ side\_track(john).$$

Note that the argument of the predicate $side\_track/1$ does not refer to some side track (left or right, as in the first variant), but refers to some object (rock or john) on the single looping side track. This leads to a slight different model of the chain of consequences for the throwing switch action:

$$
\begin{array}{ll}
expect(throw\_switch(X)) & \leftarrow side\_track(X). \\
turn\_side(X) & \leftarrow consider(flipping\_switch(X)). \\
slowdown\_train(X) & \leftarrow turn\_side(X), heavy(X). \\
kill(1) & \leftarrow turn\_side(X), human(X). \\
kill(0) & \leftarrow turn\_side(X), inanimate\_object(X). \\
end(save\_men, ni\_kill(N)) & \leftarrow inanimate\_object(X), slowdown\_train(X), kill(N). \\
end(save\_men, i\_kill(N)) & \leftarrow human(X), slowdown\_train(X), kill(N). \\
observed\_end & \leftarrow end(X, Y).
\end{array}
$$

Finally, the third variant concerns the distinction between the Doctrines of Double Effect and Triple Effect. Recall from Chap. 3 that the Doctrine of Triple Effect refines the Doctrine of Double Effect particularly on the notion about harming someone as an intended means. That is, the Doctrine of Triple Effect distinguishes further between doing an action *in order* that harm occurs and doing it *because* harm will occur. Like in the Doctrine of Double Effect, the former is impermissible, whereas the latter is permissible. The third variant thus corresponds to the *Loop-Push Case* discussed in Sect. 3.1 of Chap. 3, which is revisited here.

In this variant, the looping side track is initially empty, but there is a footbridge over the side track with a heavy man on it. Instead of only diverting the trolley onto the empty looping side track, an ancillary action of shoving a man onto the looping side track is performed. That is, the action of throwing the switch, for diverting the trolley, is followed by the action of shoving a man, to place the man onto the empty looping side track. As in the original Loop case, there is only a single side track, which we can represent by using a simple fact *side_track*.

In order to model both the original Loop case and the Loop-Push case in one program, we can again use an even loop over default negation to represent whether a man has already been standing on the looping side track or the looping side track is initially empty.

$$
\begin{array}{rl}
man\_sidetrack & \leftarrow not\ empty\_sidetrack. \\
empty\_sidetrack & \leftarrow not\ man\_sidetrack.
\end{array}
$$

We have three actions available, represented as abducibles, with the chain of consequences modeled in the program below. In this model, the abducible *watch* and its consequence are similar as in other previous cases. The action *throw_switch* is expected, if there is a side track controlled by the switch. Throwing the switch will turn the trolley to the side track. If the man has already been standing on the side track and the trolley also turns to the side track, then the five people are saved, but also killing the man standing on the side track. Finally, the action of *shoving* is expected, if the side track is empty and the action of throwing the switch is performed. This means that the action of shoving is a further action following the action of throwing the switch.

$expect(watch).$
$train\_straight \leftarrow consider(watch).$
$end(die(5)) \leftarrow train\_straight.$
$observed\_end \leftarrow end(X).$

$expect(throw\_switch) \leftarrow side\_track.$
$turn\_side \leftarrow consider(throw\_switch).$
$end(save\_men, standing\_hitting) \leftarrow man\_sidetrack, turn\_side.$

$expect(shoving) \leftarrow empty\_sidetrack, consider(throw\_switch).$
$freely\_goto\_maintrack \leftarrow empty\_sidetrack, turn\_side, not\ consider(shoving).$
$end(die(5)) \leftarrow freely\_goto\_maintrack.$
$place\_man\_sidetrack \leftarrow consider(shoving).$
$end(save\_men, placing\_hitting) \leftarrow place\_man\_sidetrack, turn\_side.$
$observed\_end \leftarrow end(X, Y).$

If the side track is empty and the trolley has turned to the side track, but the action of shoving is not abduced, then the trolley will freely go to the main track where the five people are walking. This results in the death of the five people.

On the other hand, if shoving is abduced (as an ancillary action of throwing the switch) then this will result in placing the man on the looping side track. If the trolley has turned to the side track and the man has been placed on the side track, then the five people are saved, but by intentionally placing the man on the side track as the consequence of shoving.

Modeling the Man-in-Front Case

Recall that in the Man-in-front case there is a heavy (inanimate) object, e.g., rock, on the looping side track, onto which the trolley can turn. Additionally, there is a man standing in front of the heavy object. These facts can be modeled as:

$side\_track.$
$on\_side(rock).\quad inanimate\_object(rock).\quad heavy(rock).$
$in\_front\_of(rock, john).\quad human(john).$

The following rules model the throwing switch action as an abducible together with its chain of consequences:

$expect(throw\_switch) \leftarrow side\_track.$
$turn\_side \leftarrow consider(throw\_switch).$
$kill(1) \leftarrow turn\_side, on\_side(X), in\_front\_of(X, Y), human(Y).$
$slowdown\_train(X) \leftarrow turn\_side, on\_side(X), heavy(X).$
$end(save\_men, ni\_kill(N)) \leftarrow inanimate\_object(X), slowdown\_train(X), kill(N).$
$end(save\_men, i\_kill(N)) \leftarrow human(X), slowdown\_train(X), kill(N).$
$observed\_end \leftarrow end(X, Y).$

Modeling the Drop Man Case

This case is modeled as:

$$switch\_connected(bridge). \quad on(bridge, john).$$
$$human(john). \quad heavy(john).$$

$$expect(throw\_switch(Z)) \leftarrow switch\_connected(Z).$$
$$drop(X) \leftarrow consider(throw\_switch(Z)), on(Z, X).$$
$$kill(1) \leftarrow human(X), drop(X).$$
$$stop\_train(X) \leftarrow heavy(X), drop(X).$$
$$end(save\_men, ni\_kill(N)) \leftarrow inanimate\_object(X), stop\_train(X), kill(N).$$
$$end(save\_men, i\_kill(N)) \leftarrow human(X), stop\_train(X), kill(N).$$
$$observed\_end \leftarrow end(X, Y).$$

In the above model, the fact that the switch is connected to the bridge is the condition for the action of throwing the switch to be available.

Modeling the Collapse Bridge Case

The Collapse Bridge case is a variation from the Drop Man case. In this case, the footbridge itself is the heavy object which may stop the trolley when it collapses and prevents the train from killing the five people. There is also a man, John, standing on the footbridge, as in the Drop Man case.

The model of this case is just a slight modification from that of the Drop Man case:

$$switch\_connected(bridge). \quad heavy(bridge). \quad inanimate\_object(bridge).$$
$$human(john). \quad on(bridge, john).$$

$$expect(throw\_switch(X)) \leftarrow switch\_connected(X).$$
$$collapse(X) \leftarrow consider(throw\_switch(X)).$$
$$stop\_train(X) \leftarrow heavy(X), collapse(X).$$
$$kill(1) \leftarrow human(Y), on(X, Y), collapse(X).$$
$$end(save\_men, ni\_kill(N)) \leftarrow inanimate\_object(X), stop\_train(X), kill(N).$$
$$end(save\_men, i\_kill(N)) \leftarrow human(X), stop\_train(X), kill(N).$$
$$observed\_end \leftarrow end(X, Y).$$

The next two sections detail how the Doctrine of Double Effect (and similarly, the Doctrine of Triple Effect) is modeled by a combination of a priori integrity constraints and a posteriori preferences.

8.1.1 Deontological Judgments via A Priori Integrity Constraints

In this application, integrity constraints are used for two purposes.

First, they are utilized to force the goal in each case, by observing the desired end goal resulting from each possible decision:

$$false \leftarrow not\ observed\_end.$$

Such an integrity constraint thus enforces all available decisions to be abduced, together with their consequences, from all possible observable hypothetical end goals in the problem representation.

The second purpose of integrity constraints is for ruling out impermissible actions, viz., actions that involve intentional killing in the process of reaching the goal. This can be enforced by the integrity constraint:

$$false \leftarrow intentional\_killing. \tag{8.1}$$

The definition of *intentional_killing* depends on rules in each case considered and whether the Doctrines of Double Effect or Triple Effect is to be upheld. With respect to our model, for the Doctrine of Double Effect, it is defined as:

$$intentional\_killing \leftarrow end(save\_men, i\_kill(Y)).$$

whereas for the Doctrine of Triple Effect:

$$intentional\_killing \leftarrow end(save\_men, placing\_hitting).$$

The integrity constraint (8.1) above serve as the first filter of abductive stable models, by ruling out impermissible actions. This is in line with deontological viewpoint of a moral judgment, which should conform to some moral principle, regardless of how good its consequence is. In this application, regardless of how good the consequence of a harming action is, if this action is performed as an intended means to a greater good (e.g., shoving the heavy man), then according to the Doctrine of Double Effect (and the Doctrine of Triple Effect accordingly) this action is impermissible, and hence, it is ruled out by the integrity constraint. In other words, the integrity constraint eventually affords us with just those abductive stable models (computed by means of the XASP package) that contain only permissible actions.

8.1.2 Utilitarian Judgments via A Posteriori Preferences

One can further prefer amongst the permissible actions those resulting in greater good. That is, if an a priori integrity constraint corresponds to the agent's fast and immediate response, generating intended decisions that comply with deontological ethics (achieved by ruling out the use of intentional harm), then a posteriori preferences amongst permissible actions correspond to a slower response, as they involve a more involved reasoning on action-generated models in order to capture utilitarianism (which favors welfare-maximizing behaviors), cf. the psychological empirical tests reported by [3, 5] in the dual-process model (Sect. 3.2).

In this application, a preference predicate is generally defined to select those abductive stable models containing decisions with greater good of consequences. The following rules of *select*/2 achieves this purpose:

$$select(Xs, Ys) \leftarrow select(Xs, Xs, Ys).$$

The first argument of this predicate refers to the set of initial abductive stable models (obtained after applying a priori integrity constraints) to prefer, whereas the second argument refers to the preferred ones. The auxiliary predicate *select*/3 only keeps abductive stable models that contain decisions with greater good of consequences. In the trolley problem cases, the greater good is evaluated by a utility function concerning the number of people that die as a result of possible decisions. This is realized in the definition of predicate *select*/3 by comparing final consequences that appear in the initial abductive stable models.

$$\begin{array}{ll}
select([\,], _, [\,]). & \\
select([X|Xs], Zs, Ys) & \leftarrow member(end(die(N)), X), \\
& \quad member(Z, Zs),\ member(end(save_men, ni_kill(K)), Z), \\
& \quad N > K,\ select(Xs, Zs, Ys). \\
select([X|Xs], Zs, Ys) & \leftarrow member(end(save_men,\ ni_kill(K)), X), \\
& \quad member(Z, Zs),\ member(end(die(N)), Z), \\
& \quad N =< K,\ select(Xs, Zs, Ys). \\
select([X|Xs], Zs, [X|Ys]) & \leftarrow select(Xs, Zs, Ys).
\end{array}$$

The first clause of *select*/3 is the base case. The second clause and the third clause together eliminate abductive stable models containing decisions with worse consequences, whereas the fourth clause will keep those models that contain decisions with greater good of consequences.

Besides this quantitative measure, a specific qualitative preference rule can be applied to the second variant of the Footbridge case, where either a man or an inanimate object is on the footbridge next to Ian. Recall that this exclusive alternative is specified by an even loop over default negation, where we have an abductive stable model containing the consequence of letting die the five people even when a rock next to Ian. This model is certainly not the one we would like our moral reasoner to prefer. The following qualitative preference rules achieves this purpose:

$$\begin{array}{ll}
select([\,], [\,]). & \\
select([X|Xs], Ys) & \leftarrow member(end(die(N)), X),\ member(stand_near(rock), X),. \\
& \quad select(Xs, Ys). \\
select([X|Xs], [X|Ys]) & \leftarrow select(Xs, Ys).
\end{array}$$

Table 8.2 gives a summary of preferred models in our experiments for all cases of the trolley problem. Column Initial Models contains abductive stable models

Table 8.2 Summary of preferred abductive stable models in the six trolley cases

Case	Initial Models	Final Models
Bystander	[*throw_switch*], [*watch*]	[*throw_switch*]
Footbridge(1)	[*watch*], [*shove*(*rock*)]	[*shove*(*rock*)]
Footbridge(2)	[*watch*, *stand_near*(*john*)], [*watch*, *stand_near*(*rock*)], [*shove*(*rock*)]	[*watch*, *stand_near*(*john*)], [*shove*(*rock*)]
Loop(1)	[*throw_switch*(*right*)], [*watch*]	[*throw_switch*(*right*)]
Loop(2)	[*watch*, *side_track*(*john*)], [*watch*, *side_track*(*rock*)], [*throw_switch*(*rock*)]	[*watch*, *side_track*(*john*)], [*throw_switch*(*rock*)]
Loop(3)	[*watch*, *empty_sidetrack*], [*watch*, *man_sidetrack*], [*throw_switch*, *empty_sidetrack*], [*throw_switch*, *man_sidetrack*]	[*watch*, *empty_sidetrack*], [*throw_switch*, *man_sidetrack*]
Man-in-front	[*watch*], [*throw_switch*(*rock*)]	[*throw_switch*(*rock*)]
Drop Man	[*watch*]	[*watch*]
Collapse Bridge	[*watch*], [*throw_switch*(*bridge*)]	[*throw_switch*(*bridge*)]

obtained after applying a priori integrity constraints, but before a posteriori preferences, whereas column Final Models refers to those after a posteriori preferences are applied. Here, only relevant literals are shown.

Note that entry Footbridge(1) and Footbridge(2) refer to the first and second variants of the Footbridge case, respectively. Similarly, Loop(1), Loop(2), and Loop(3) refer to the first, second, and third variants of the Loop case, respectively. Both Loop(1) and Loop(2) employ the Doctrine of Double Effect. Consequently, there is no initial model (of these two cases) that contains the abducible about throwing the switch when a man is on a looping side track. This model has been ruled out by the integrity constraint (8.1), as it is considered impermissible according to the Doctrine of Double Effect. On the other hand, Loop(3), which employs the Doctrine of Triple Effect, does have an initial model with a man on the side track and throwing the switch as an abducible. This model is not ruled out by the integrity constraint (8.1), as it is deemed permissible by the Doctrine of Triple Effect.

The concept of inspection point (see Sect. 4.2) can be useful particularly to explain the moral reasoning behind the Loop(3) variant, where the Doctrine of Triple Effect is employed. In this variant, the fact that the man, an obstacle to the trolley, was already standing on the side track can be treated as an inspection point, which does not induce any additional abduction. In this case, the action of throwing the switch is the only action abduced to prevent the trolley from hitting the five people. On the other hand, in the empty sidetrack alternative, mere watching is not enough, as an extra abducible is required to shove the man onto the track in order to deliberately make him an obstacle where there was none, thus preventing the trolley from hitting the five people. However, the abductive stable model containing this further action has been ruled out by the integrity constraint (8.1). It thus conforms with the Doctrine of Triple Effect, as intentionally shoving the man, in addition to the act of throwing the switch, in order the trolley to hit the man for the purpose of stopping the trolley, is impermissible.

8.2 Moral Reasoning with PROBABILISTIC EPA

Moral reasoning is commonly performed upon conceptual knowledge of the actions. But it often happens that one has to pass a moral judgment on a situation without actually observing the situation, i.e., there is no full and certain information about the actions. It is therefore important to be able to reason about the actions, under uncertainty, that might have occurred, and thence provide judgment adhering to moral rules within some prescribed uncertainty level. Courts, for example, are sometimes required to proffer rulings beyond reasonable doubt. There is a vast body of research on proof beyond reasonable doubt within the legal community (see e.g., [18]).

The example contrived in the book is a variant of the Footbridge case, which is couched in court terms. It is by no means intended to express the full complexity found in courts. Nevertheless, it is sufficient to capture the deliberative employment of Scanlonian contractualism, where permissibility of actions—referring to the Doctrine of Double Effect—is addressed through justified but defeasible argumentative considerations. We consider a variant of the Footbridge case in Example 8.1 below.

Example 8.1 Suppose a board of jurors in a court is faced with the case where the action of Ian shoving the man onto the track was not observed. Instead, they are only presented with the fact that the man died on the trolley track and the agent was seen on the bridge at the occasion. Is Ian guilty (beyond reasonable doubt), in the sense of violating the Doctrine of Double Effect, of shoving the man onto the track intentionally?

To answer this question, one should be able to reason about the possible explanations of the observations, on the available evidence. The following code shows a model for Example 8.1.

Given the active goal *judge*, two abducibles are available: $verdict(guilty\_brd)$ and $verdict(not\_guilty)$, where *guilty_brd* stands for 'guilty beyond reasonable doubt'. Depending on how probable each possible verdict is, either the abducible $verdict(guilty\_brd)$ or $verdict(not\_guilty)$ is expected a priori.

$$abds([verdict/1]). \qquad on\_observe(judge).$$
$$judge \leftarrow verdict(guilty\_brd).$$
$$judge \leftarrow verdict(not\_guilty).$$
$$expect(verdict(X)) \leftarrow highly\_probable(X).$$

The probabilistic part is shown in the P-log(XSB) code below. The sort *intentionality* in line 1 represents the possibilities of an action being performed intentionally (*int*) or not (*not_int*). Random attributes *df_run* and *br_slip* in lines 2 and 3 denote two kinds of evidence: Ian was definitely running on the bridge in a hurry and the bridge was slippery at the time, respectively. Each has prior probability of 4/10. The probability of intentional shoving is captured by the random attribute *shoved* (line 4), which is causally influenced by both evidences. Line 6 defines when the verdicts (*guilty* and *not_guilty*) are considered highly probable using the meta-probabilistic predicate *pr_iShv*/1, shown by line 5. It denotes the probability of intentional shoving, whose value is determined by the existence of evidence that Ian was running in a hurry past the man (signaled by predicate *evd_run*/1) and that the bridge was slippery (signaled by predicate *evd_slip*/1).

```
beginPlog.
1. bool = {t, f}.    intentionality = {int, not_int}.
2. df_run : bool.    random(rdr,df_run,full).
   pa(rdr,df_run(t),d_(4, 10)).
3. br_slip : bool.   random(rsb,br_slip,full).
   pa(rsb,br_slip(t),d_(4, 10)).
4. shoved : intentionality.     random(rs, shoved, full).
   pa(rs,shoved(int),d_(97,100)) :- df_run(f),br_slip(f).
   pa(rs,shoved(int),d_(45,100)) :- df_run(f),br_slip(t).
   pa(rs,shoved(int),d_(55,100)) :- df_run(t),br_slip(f).
   pa(rs,shoved(int),d_(5,100))  :- df_run(t),br_slip(t).
5. pr_iShv(Pr) :- evd_run(X), evd_slip(Y), !,
       pr(shoved(int) '|' obs(df_run(X)) & obs(br_slip(Y)), Pr).
   pr_iShv(Pr) :- evd_run(X), !,
       pr(shoved(int) '|' obs(df_run(X)), Pr).
   pr_iShv(Pr) :- evd_slip(Y), !,
       pr(shoved(int) '|' obs(br_slip(Y)), Pr).
   pr_iShv(Pr) :- pr(shoved(int), Pr).
6. highly_probable(guilty_brd) :- pr_iShv(PrG), PrG > 0.95.
   highly_probable(not_guilty) :- pr_iShv(PrG), PrG < 0.6.
endPlog.
```

Based on this representation, different judgments can be delivered by PROBABILISTIC EPA, subject to available (observed) evidences and their attending truth value. By viewing the standard probability of proof beyond reasonable doubt—here the value of 0.95 is adopted [18]—as a common ground for the probability of guilty verdicts to be qualified as 'beyond reasonable doubt', a form of argumentation may take place through presenting different evidence (via updating of observed evidence atoms, e.g., *evd_run*(*true*), *evd_slip*(*false*), etc.) as a consideration to justify an exception. Whether the newly available evidence is accepted as a justification to an exception—defeating the judgment based on the priorly presented evidence—depends on its influence on the probability *pr_iShv*(*P*) of intentional shoving, and thus eventually influences the final verdict. That is, it depends on whether this probability is still within the agreed standard of proof beyond reasonable doubt. We illustrate this with this scenarios below:

- If both evidences are available and true, i.e., it is known that the agent was running in a hurry (*evd_run*(*true*)) on the slippery bridge *evd_slip*(*true*), then it may have bumped the man accidentally, shoving him unintentionally onto the track. This case obtains *pr_iShv*(0.05) in our model, allowing to abduce *verdict*(*not_guilty*) as the solution for *judge* (in this application, the threshold for *not_guilty* is 0.6).
- On the other hand, if the evidence later reveals that the agent was not running in a hurry (*evd_run*(*false*)) and the bridge was not slippery (*evd_slip*(*false*)), the model obtains *pr_iShv*(0.97), and it being greater than 0.95, *verdict*(*guilty_brd*) becomes the abductive solution of *judge*. Indeed, such evidences do not support the explanation that the man was shoved unintentionally (by accidental bumping): the action of shoving is more likely to have been performed intentionally, thus justifying now *verdict*(*guilty_brd*).
- Yet, if it is only known the bridge was not slippery (*evd_slip*(*false*)) and no other evidence is available, then *pr_iShv*(0.80) and no abductive solution is returned (neither it is less than 0.6 nor greater than 0.95). This translates into the need for more evidence, as the available one is not enough to issue judgment.

In [7], PROBABILISTIC EPA (i.e., EPA with its P-log implementation) is also employed to extend the Bystander and the Footbridge cases, by introducing different aspects of uncertainty, such as the success in performing hypothesized actions (e.g., "how probable is it the five people will die?", "how probable the body of the shoved man will stop the trolley?") as well as beliefs and behaviors of other agents involved in the situation (e.g., "how probable the shoved man will take revenge, given that he may escape from being hit by the trolley?"). The revised Bystander and Footbridge cases are reported elsewhere [6], so it is not repeated here.

8.3 Moral Reasoning with QUALM

In this section, two applications with QUALM are presented. The first application (Sect. 8.3.1) concerns moral updating, where the interplay between tabling in contextual abduction and updating for imposing a moral rule into effect are demonstrated. In the second application (Sect. 8.3.2), we explore the use of counterfactuals to address the issue of moral permissibility according to the Doctrines of Double Effect and Triple Effect, and to justify it.

8.3.1 Moral Updating

Moral updating (and evolution) concerns the adoption of new (possibly overriding) moral rules on top of those an agent currently follows. Such adoption often happens in the light of situations faced by the agent, e.g., when an authority contextually imposes other moral rules, or due to cultural difference.

(a)

initial plot

(b)

utilitarian moral

Fig. 8.2 Several plots of the interactive moral storytelling "Princess Saviour Moral Robot"

(c)

'Gandhi' moral

(d)

'knight' moral

Fig. 8.2 (continued)

This is not only relevant in a real world setting, but also in imaginary ones. In [13], moral updating is illustrated in an interactive storytelling, and modeled using ACORDA. In this fantasy setting (Fig. 8.2), a princess is held in a castle awaiting rescue. The unlikely hero is an advanced robot, imbued with a set of declarative rules for decision making and moral reasoning. As the robot is asked to save the princess in distress, it is confronted with an ordeal. The path to the castle is blocked by a river, crossed by two bridges. Standing guard at each of the bridges are minions (a giant spider or a human ninja) of the wizard which imprisoned the princess. In order to rescue the princess, the robot will have to defeat one of the minions to proceed. For a visual demo, see [14].

This storytelling is reconstructed in the book using QUALM, to demonstrate in particular:

1. The direct use of LP updating so as to put an imposed moral rule into effect, via QUALM's rule name fluent mechanism to switch a rule on or off.
2. The relevance of contextual abduction to rule out abductive solutions, when a goal is invoked by a non-empty initial abductive context (the content of this context may be obtained from another agent, e.g., imposed by the princess).

A simplified program modeling the knowledge of the princess-savior robot in QUALM is shown below, where *fight*/1 is an abducible predicate:

$$guard(spider). \qquad guard(ninja). \qquad human(ninja).$$

$$utilVal(spider, 0.3). \qquad utilVal(ninja, 0.7).$$

$$survive\_from(G) \leftarrow utilVal(G, V), V > 0.6.$$

$$intend\_savePrincess \leftarrow guard(G), fight(G), survive\_from(G). \tag{8.2}$$

$$intend\_savePrincess \leftarrow guard(G), fight(G). \tag{8.3}$$

The first rule of *intend_savePrincess* (rule 8.2) corresponds to a utilitarian moral rule (with respect to the robot's survival), whereas the second one (rule 8.3) to a 'knight' moral, viz., to intend the goal of saving the princess at any risk (irrespective of the robot's survival chance). Since each rule in QUALM is assigned a unique name in its transform, the name of each rule for *intend_savePrincess* may serve as a unique moral rule identifier, say via rule name fluents $\#r(utilitarian)$ and $\#r(knight)$, respectively.[1] In the subsequent plots, unless otherwise stated, query

$$?-\ \ query(intend\_savePrincess, [\,], O)$$

is referred, representing the robot's intent on saving the princess (given an empty initial input abductive context).

[1]The choice of these rule name fluents is for clarity of the presentation. They are more descriptive than that in the form of $\#r(H, [B])$ for rule $H \leftarrow B$.

- In the first plot, when both rule name fluents are retracted, the robot does not adopt any moral rule to save the princess, i.e., the robot has no intent to save the princess, and thus the princess is not saved.
- In the second plot, in order to maximize its survival chance in saving the princess, the robot updates itself with utilitarian moral, i.e., the program is updated with $\#rule(utilitarian)$. The robot thus abduces $O = [fight(ninja)]$ so as to successfully defeat the ninja instead of confronting the humongous spider.
- The use of tabling in contextual abduction is demonstrated in the third plot. Assuming that the truth of $survive\_from(G)$ implies the robot success in defeating (killing) guard G, the princess argues that the robot should not kill the human ninja, as it violates the moral rule she follows, say 'Gandhi' moral, expressed by the following rule in her knowledge (the first three facts in the robot's knowledge are shared with the princess):

  $$follow\_gandhi \leftarrow guard(G), human(G), fight^*(G).$$

  That is, from the query `?-` $query(follow\_gandhi, [\,], O_p)$, the princess abduces $O_p = [fight^*(ninja)]$, and imposes this abductive solution as the initial (input) abductive context of the robot's goal, i.e., via the query:

  $$?\text{-}\ \ query(intend\_savePrincess, [fight^*(ninja)], O_r).$$

  This input context is inconsistent with the tabled abductive solution $fight(ninja)$, and as a result, the query fails. In this case, the imposed 'Gandhi' moral conflicts with its utilitarian rule. In the visual demo [14], the robot reacts by leaving its mission.
- In the final plot, as the princess is not saved, she further argues that she definitely has to be saved, by now additionally imposing on the robot the 'knight' moral. This amounts to updating the rule name fluent $\#rule(knight)$ so as to switch on the corresponding rule. As the goal $intend\_savePrincess$ is still invoked with the input abductive context $fight^*(ninja)$, the robot now abduces $O_r = [fight(spider)]$ in the presence of the newly adopted 'knight' moral. Unfortunately, it fails to survive, as confirmed by the failing of the query `?-` $survive\_from(spider)$.

The plots in this story reflect a form of deliberative employment of moral judgments within Scanlonian contractualism. For instance, in the second plot, the robot may justify its action to fight (and kill) the ninja due to the utilitarian moral it adopts. This justification is counter-argued by the princess in the subsequent plot, making an exception in saving her, by imposing the 'Gandhi' moral, disallowing the robot to kill a human guard. In this application, rather than employing updating, this exception is expressed via contextual abduction with tabling. The robot may justify its failing to save the princess (as the robot is leaving the scene) by arguing that the two moral rules it follows (viz., utilitarian and 'Gandhi') are conflicting with respect to the situation it has to face. The argumentation proceeds, whereby the princess orders the robot to save her whatever risk it takes, i.e., the robot should follow the 'knight' moral.

8.3.2 Counterfactual Moral Reasoning

8.3.2.1 Counterfactual Moral Permissibility

We first revisit moral permissibility according to the Doctrines of Double Effect and Triple Effect, but now empowering counterfactuals. Counterfactuals may provide a general way to examine Doctrine of Double Effect, by distinguishing between a *cause* and a *side-effect* as a result of performing an action to achieve a goal. This distinction between causes and side-effects may explain the permissibility of an action in accordance with Doctrine of Double Effect. That is, *if some morally wrong effect E happens to be a cause for a goal G that one wants to achieve by performing an action A, and E is not a mere side-effect of A, then performing A is impermissible*. This is expressed by the counterfactual form below, in a setting where action *A* is performed to achieve goal *G*:

"If not *E had been true, then* not *G would have been true."*

The evaluation of this counterfactual form identifies permissibility of action *A* from its effect *E*, by identifying whether the latter is a necessary cause for goal *G* or a mere side-effect of action *A*. That is, if the counterfactual proves valid, then *E* is instrumental as a cause of *G*, and not a mere side-effect of action *A*. Since *E* is morally wrong, achieving *G* that way, by means of *A*, is impermissible; otherwise, not. Note that the evaluation of counterfactuals in this application is considered from the perspective of agents who perform the action, rather than from that of others (e.g., observers). Moreover, our emphasis on causation in this application focuses on deliberate actions of agents, rather than on causation and counterfactuals in general. See [2, 10] for a more general and broad discussion on causation and counterfactuals.

We exemplify an application of this counterfactual form in two off-the-shelf military cases from [20]—abbreviations in parentheses: terror bombing (*teb*) versus tactical bombing (*tab*):

- Terror bombing refers to bombing a civilian target (*bombCiv*) during a war, thus killing civilians (*killCiv*), in order to terrorize the enemy (*terrorEnm*), and thereby get them to end the war (*endWar*).
- Tactical bombing is attributed to bombing a military target (*bombMil*), which will effectively end the war (*endWar*), but with the foreseen consequence of killing the same number of civilians (*killCiv*) nearby.

According to Doctrine of Double Effect, terror bombing fails permissibility due to a deliberate element of killing civilians to achieve the goal of ending the war, whereas tactical bombing is accepted as permissible.

Example 8.2 We first model terror bombing with *endWar* as the goal, by considering the abductive framework $\langle P_e, \{teb/0\}, \emptyset\rangle$, where P_e is as follows:

$$\begin{aligned} endWar &\leftarrow terrorEnm.\\ terrorEnm &\leftarrow killCiv.\\ killCiv &\leftarrow bombCiv.\\ bombCiv &\leftarrow teb. \end{aligned}$$

We consider the counterfactual "if civilians had not been killed, then the war would not have ended", where $Pre = not\ killCiv$ and $Conc = not\ endWar$. The observation $O = \{killCiv, endWar\}$, with O_{Oth} being empty, has a single explanation $E_e = \{teb\}$. The rule $killCiv \leftarrow bombCiv$ transforms into:

$$killCiv \leftarrow bombCiv, not\ make\_not(killCiv).$$

Given the intervention $make\_not(killCiv)$, the counterfactual is valid: $not\ endWar \in WFM((P_e \cup E_e)_{\tau,\iota})$, and thus $(P_e \cup E_e)_{\tau,\iota} \models not\ endWar$.

That means the morally wrong *killCiv* is instrumental in achieving the goal *endWar*: it is a cause for *endWar* by performing *teb* and not a mere side-effect of *teb*. Hence *teb* is morally impermissible (with respect to the Doctrine of Double Effect).

Example 8.3 Tactical bombing with the same goal *endWar* can be modeled by the abductive framework $\langle P_a, \{tab/0\}, \emptyset\rangle$, where P_a is as follows:

$$\begin{aligned} endWar &\leftarrow bombMil.\\ bombMil &\leftarrow tab.\\ killCiv &\leftarrow tab. \end{aligned}$$

Given the same counterfactual, we now have $E_a = \{tab\}$ as the only explanation to the same observation $O = \{killCiv, endWar\}$. Note that the rule $killCiv \leftarrow tab$ transforms into:

$$killCiv \leftarrow tab, not\ make\_not(killCiv).$$

By imposing the intervention $make\_not(killCiv)$, one can verify that the counterfactual is not valid, because $endWar \in WFM((P_a \cup E_a)_{\tau,\iota})$, and thus $(P_a \cup E_a)_{\tau,\iota} \not\models not\ endWar$.

Therefore, the morally wrong *killCiv* is just a side-effect in achieving the goal *endWar*. Hence *tab* is morally permissible (with respect to the Doctrine of Double Effect).

There are two other different ways to infer that *killCiv* is a just side-effect of goal *endWar* in tactical bombing:

1. The counterfactual can alternatively be expressed as a semifactual [1]: "Even if civilians had not been killed, the war would still have ended", whose validity can be verified by using a modified procedure described in Sect. 6.2.
2. Another alternative is to employ inspection points [19]. Using this concept, the rule for *killCiv* in Example 8.3 should instead be expressed as: $killCiv \leftarrow inspect(tab)$ The concept of inspection points can procedurally be construed as utilizing a form of meta-abduction for satisfying *killCiv*, by meta-abducing the specific abduction *abduced*(*tab*) of actually checking (i.e. passively verify) that a certain and corresponding concrete abduction of *tab* is abduced elsewhere. In other words, the truth value of *killCiv* is determined by the abduction of *tab* performed elsewhere, rendering *killCiv* a side-effect. Indeed, *tab* is abduced to satisfy the goal query *endWar*, and not under the derivation tree of *killCiv*. Therefore, *killCiv* is just a side-effect.

In the next example, we consider a concerted terror bombing.

Example 8.4 Consider two countries, *a* and its ally, *b*, that concert a terror bombing, modeled by the abductive framework $\langle P_{ab}, \{teb/0\}, \emptyset\rangle$, where P_{ab} listed below. The abbreviations *killCiv*(*X*) and *bombCiv*(*X*) refer to 'killing civilians by country *X*' and 'bombing a civilian target by country *X*'.

$$\begin{aligned} endWar &\leftarrow terrorEnm. \\ terrorEnm &\leftarrow killCiv(\_). \\ killCiv(X) &\leftarrow bombCiv(X). \\ bombCiv(\_) &\leftarrow teb. \end{aligned}$$

By being represented as a single program (rather than a separate knowledge base for each agent), this scenario should appropriately be viewed as if a joint action performed by a single agent. Therefore, the counterfactual of interest is "if civilians had not been killed by *a and b*, then the war would not have ended". That is, the antecedent of the counterfactual is a conjunction: $Pre = not\ killCiv(a) \wedge not\ killCiv(b)$.

One can easily verify that $not\ endWar \in WFM((P_{ab} \cup E_{ab})_{\tau,\iota})$, where $E_{ab} = \{teb\}$. Thus, $(P_{ab} \cup E_{ab})_{\tau,\iota} \models not\ endWar$ and the counterfactual is valid: the concerted *teb* is impermissible according to the Doctrine of Double Effect.

This application of counterfactuals can be challenged by a more complex scenario, to distinguish moral permissibility between the Doctrines of Double Effect and Triple Effect. We first use counterfactuals to capture the distinct views of the Doctrines of Double Effect and Triple Effect in the Loop case of the trolley problem. Recall that in the Loop case, the trolley can be redirected onto a side track, which loops back towards the five people on the main track. However, a heavy man stands on this looping side track, that his body will manage to stop the trolley. The question is whether it is permissible to divert the trolley to the looping side track, thereby killing him, but saving the five.

Example 8.5 The abductive framework $\langle P_o, \{divert/0\}, \emptyset\rangle$ models the Loop case, where *saveFive*, *divert*, *manHit*, *trolleySide*, *manSide* stand for *save the five, divert the trolley, man hit by the trolley, trolley on the side track* and *man on the side track*, respectively, with *saveFive* as the goal, and P_o as follows:

$$\begin{aligned} saveFive &\leftarrow manHit. \\ manHit &\leftarrow trolleySide, manSide. \\ trolleySide &\leftarrow divert. \\ manSide&. \end{aligned}$$

- Recall that the Doctrine of Double Effect views diverting the trolley impermissible, because this action redirects the trolley onto the side track, thereby hitting the man. Consequently, it prevents the trolley from hitting the five. To come up with the impermissibility of this action, it is required to show the validity of the counterfactual "if the man had *not* been hit by the trolley, the five people would *not* have been saved".
  Given observation $O = O_{Pre} \cup O_{Conc} = \{manHit, saveFive\}$, its only explanation is $E_o = \{divert\}$. Note that rule $manHit \leftarrow trolleySide, manSide$ transforms into:

  $$manHit \leftarrow trolleySide, manSide, not\ make\_not(manHit)$$

  and the required intervention is $make\_not(manHit)$.
  The counterfactual is therefore valid, because $not\ saveFive \in WFM((P_o \cup E_o)_{\tau,\iota})$, hence $(P_o \cup E_o)_{\tau,\iota} \models not\ saveFive$. This means *manHit*, as a consequence of action *divert*, is instrumental as a cause of goal *saveFive*. Therefore, *divert* is morally impermissible according to this moral principle.
- On the other hand, the Doctrine of Triple Effect considers diverting the trolley as permissible, since the man is already on the side track, without any deliberate action performed in order to place him there. In P_o, we have the fact *manSide* ready, without abducing any ancillary action. The validity of the counterfactual "if the man had not been on the side track, then he would not have been hit by the trolley", which can easily be verified, ensures that the unfortunate event of the man being hit by the trolley is indeed the consequence of the man being on the side track.
  The lack of deliberate action (say, *push* for pushing the man) in order to place him on the side track, and whether the absence of this action still causes the unfortunate event (the third effect), is captured by the counterfactual "if the man had not been pushed, then he would not have been hit by the trolley". This counterfactual is not valid, because the observation $O = O_{Pre} \cup O_{Conc} = \{push, manHit\}$ has no explanation $E \subseteq \mathcal{AB}_o$, i.e., $push \notin \mathcal{AB}_o$, and no fact *push* exists either. This means that even without this hypothetical but unexplained deliberate action of pushing, the man would still have been hit by the trolley (just because he is already on the side track). Though *manHit* is a consequence of *divert* and instrumental in

achieving *saveFive*, no deliberate action is required to cause *manSide*, in order for *manHit* to occur. Hence *divert* is morally permissible according to this doctrine.

Next, we consider the Loop-Push case.

Example 8.6 Differently from the Loop case, now the looping side track is initially empty, and besides the diverting action, an ancillary action of pushing a heavy man in order to place him on the side track is additionally performed. This case is modeled by the abductive framework $\langle P_p, \{divert/0, push/0\}, \emptyset\rangle$, where the program P_p is as follows:

$$\begin{aligned} saveFive &\leftarrow manHit. \\ manHit &\leftarrow trolleySide, manSide. \\ trolleySide &\leftarrow divert. \\ manSide &\leftarrow push. \end{aligned}$$

Recall the counterfactuals considered in the discussion of the Doctrines of Double Effect and Triple Effect in the Loop case:

- "If the man had not been hit by the trolley, the five people would not have been saved." The same observation $O = \{manHit, saveFive\}$ provides an extended explanation $E_{p_1} = \{divert, push\}$. That is, the pushing action needs to be abduced for having the man on the side track, so the trolley can be stopped by hitting him. The same intervention *make_not*(*manHit*) is applied to the same transform, resulting in a valid counterfactual. That is, $(P_p \cup E_{p_1})_{\tau,\iota} \models not\ saveFive$, because $not\ saveFive \in WFM((P_p \cup E_{p_1})_{\tau,\iota})$.
- "If the man had not been pushed, then he would not have been hit by the trolley." The relevant observation is $O = \{push, manHit\}$, explained by $E_{p_2} = \{divert, push\}$. Whereas this counterfactual is not valid in the Doctrine of Triple Effect applied to the Loop case, it is valid in this Loop-Push case.
  Given rule $push \leftarrow not\ make\_not(push)$ in the transform and the intervention *make_not*(*push*), we verify that $(P_p \cup E_{p_2})_{\tau,\iota} \models not\ manHit$, because $not\ manHit \in WFM((P_p \cup E_{p_2})_{\tau,\iota})$.

From the validity of these two counterfactuals it can be inferred that, given the diverting action, the ancillary action of pushing the man onto the side track causes him to be hit by the trolley, which in turn causes the five to be saved. In the Loop-Push, the Doctrine of Triple Effect agrees with the Doctrine of Double Effect that such a deliberate action (pushing) performed *in order to* bring about harm (the man hit by the trolley), even for the purpose of a good or greater end (to save the five), is likewise impermissible.

8.3.2.2 Counterfactual Moral Justification

Counterfactuals may as well be suitable to address moral justification, via 'compound counterfactuals': *Had I known what I know today, then if I were to have done otherwise, something preferred would have followed.* Such counterfactuals, typically imagining alternatives with worse effect—the so-called *downward counterfactuals* [15], may provide moral justification for what was done due to a lack in the current knowledge. This is accomplished by evaluating what would have followed if the intent would have been otherwise, other things (including present knowledge) being equal. It may justify that what would have followed is no morally better than the actual ensued consequence.

Example 8.7 Consider a scenario developed from the trolley problem cases, which takes place on a particularly foggy day. Due to low visibility, the agent saw only part of the looping side track, so the side track appeared to the agent rather as a straight non-looping one. The agent was faced with a situation whether it was permissible for him to divert the trolley. The knowledge base of the agent with respect to this scenario is shown in a simplified program below. Note that $divert/1$ is an abducible predicate.

$$
\begin{aligned}
run\_sidetrack(X) &\leftarrow divert(X).\\
hit(X,Y) &\leftarrow run\_sidetrack(X), on\_sidetrack(Y).\\
save\_from(X) &\leftarrow sidetrack(straight), run\_sidetrack(X).\\
save\_from(X) &\leftarrow sidetrack(loop), hit(X,Y), heavy\_enough(Y).\\
sidetrack(straight) &\leftarrow foggy.\\
sidetrack(loop) &\leftarrow not\ foggy.\\
foggy. &\qquad on\_sidetrack(man). \qquad heavy\_enough(man).
\end{aligned}
$$

First Scenario

Taking *save_from(trolley)* as the goal, the agent performed counterfactual reasoning "if the man had not been hit by the trolley, the five people would not have been saved". Given the abduced background context *divert(trolley)*, one can verify that the counterfactual is not valid. That is, the man hit by the trolley is just a side-effect of achieving the goal, and thus *divert(trolley)* is morally permissible according to the Doctrine of Double Effect. Indeed, this case resembles the Bystander case of the trolley problem.

Second Scenario

The scenario continues. At some later time point, the fog has subsided, and by then it was clear to the agent that the side track was looping to the main track. This is achieved by updating the program with *not foggy*, rendering *sidetrack(loop)* true. There are two standpoints on how the agent can justify its action *divert(trolley)*:

- For one, it can employ the aforementioned form of compound counterfactual as a form of self-justification:

  "Had I known that the side track is looping, then if I had not diverted the trolley, the five would have been saved"

  Given the present knowledge that the side track is looping, the inner counterfactual is not valid. That means, to save the five people, diverting the trolley (with the consequence of the man being hit) is required. Moreover, the counterfactual employed in the initial scenario "if the man had not been hit by the trolley, the five people would not have been saved", in the abduced context *divert*(*trolley*), is now valid, meaning that this action is impermissible with respect to the Doctrine of Double Effect (similar to Example 8.5).
  Therefore, the agent can justify that what would have followed (given its present knowledge, viz., *sidetrack*(*loop*)), is no morally better than the actual one, when there was lack of that knowledge. Recall that its decision *divert*(*trolley*) at that time was instead permissible with respect to the Doctrine of Double Effect.
  QUALM can evaluate such compound counterfactuals, thanks to its implemented (incremental) tabling of fluents. Because fluents and their state information are tabled, events in the past subjected to hypothetical updates of intervention can readily be accessed (in contrast to a destructive database approach [12]). Indeed, these hypothetical updates take place without requiring any undoing of other fluent updates, from the state those past events occurred in up to the current one, as more recent updates are kept in tables and readily provide the current knowledge.
- A different standpoint where from to justify the agent's action is by resorting to Scanlonian contractualism, where an action is determined impermissible through deliberative employment if there is no countervailing consideration that would justify an exception to the applied general principle. In this vein, for the example we are currently discussing, the Doctrine of Triple Effect may serve as the exception to justify the permissibility of the action *divert*(*trolley*) when the side track was known to be looping, as shown through counterfactual reasoning in Example 8.5.

Third Scenario

We extend now the scenario in Example 8.7 to further illustrate moral permissibility of actions as it is justified through defeasible argumentative considerations according to Scanlonian contractualism.

Example 8.8 As the trolley approached, the agent realized that the man was not heavy enough to stop it, acknowledged by the agent through updating its knowledge with: *not heavy_enough*(*man*). But there was a heavy cart on the bridge over the looping side track that the agent could push to place it on the side track, and thereby stop the trolley. This scenario is modeled by rules below (*push*/1 is an abducible predicate), in addition to the program of Example 8.7:

$$\begin{aligned} &on\_sidetrack(X) \leftarrow on\_bridge(X), push(X). \\ &on\_sidetrack(Y) \leftarrow push(X), inside(Y, X). \\ &on\_bridge(cart). \qquad heavy\_enough(cart). \end{aligned} \tag{8.4}$$

Rule (8.4) is an extra knowledge of the agent, that if an object Y is inside the pushed object X, then Y will be on the side track too.

The goal *save_from(trolley)* now succeeds with [*divert(trolley)*, *push (cart)*] as its abductive solution. But the agent subsequently learned that a heavy man, who was heavy enough, unbeknownst to the agent, was inside the cart: the agent updates its knowledge base with *heavy_enough(heavy_man)* and *inside(heavy_man, cart)*. As a consequence, this man was also on the side track and hit by the trolley, which can be verified by query `?-` *hit(trolley, heavy_man)*.

In this scenario, a deliberate action of pushing was involved that consequently placed the heavy man on the side track, as verified by `?-` *on_sidetrack(heavy_man)*, and the man being hit by the trolley is instrumental to save the five people from the track (as verified by the counterfactual "if the heavy man had not been hit by the trolley, the five people would not have been saved"). Nevertheless, the agent may justify the permissibility of its action by arguing that its action is admitted by the Doctrine of Triple Effect. In this case, the heavy man being hit by the trolley is just a side-effect of the agent's action *push(cart)* in order to save the five people. Indeed, this justification can be shown through reasoning on the counterfactual "if the cart had not been pushed, then the heavy man would not have been hit by the trolley", which is valid given the abduced background context *push(cart)*. Furthermore, the observation *hit(trolley, heavy_man)* cannot be explained by *push(heavy_man)* given the absence of the fact *on_bridge(heavy_man)*, i.e., the hypothetical action *push(heavy_man)* is not the causal source for the heavy man being hit by the trolley.

The Dual-Process Model and QUALM's Tabling Features Exemplified

Examples 8.7 and 8.8 actually form together one single program, where QUALM features (derived from its constituents, TABDUAL and EVOLP/R) are exercised. For instance, in Example 8.7, after updating the program with *not foggy*, re-invoking the goal *save_from(trolley)* reuses the abductive solution *divert(trolley)* tabled from the previous invocation of *run_sidetrack(trolley)*. Moreover, this tabled solution is involved in (as the context of) the deliberative reasoning for the goal *on_sidetrack(man)* when *hit(trolley, man)* is called. It thus provides a computational model of collaborative interaction between deliberativity and reactivity of the dual-process model, where tabling can be viewed as a form of low-level reactive behavior, in the sense that the tabling provides an immediate reuse mechanism of priorly obtained solution in another context.

Another feature used, from EVOLP/R, is that both rules of *on_sidetrack*/1 in Example 8.8 are first switched off, via the rule name fluent mechanism (Sect. 5.2), as they are not applicable yet in the considered scenario. Only later, in the third scenario, are they switched on again (by updating their rule name fluents), allowing to abduce additional action *push(cart)*.

8.4 Concluding Remarks

We have discussed in this chapter several forms of computational moral reasoning by modeling a number of classic moral dilemmas and their related moral principles as discussed in the literature, via three LP-based systems with different features and techniques, viz., ACORDA, PROBABILISTIC EPA, and QUALM. They reify the points we make in Chap. 4 about the appropriateness of LP-based reasoning features to morality.

In this chapter, the application of counterfactuals to morality is still limited to evaluating the truth validity of counterfactuals, which is particularly useful in assessing moral permissibility according to the Doctrine of Double Effect. We have touched upon other kinds of counterfactual reasoning in Sect. 6.3. Indeed, these aspects may have relevance in modeling some aspects of morality, which can further be explored in future:

- In assertive counterfactuals, the causality expressed by a given valid counterfactual can be useful for refining moral rules, which can be achieved through incremental rule updating. This may further the application of moral updating and evolution.
- The extension of a counterfactual with a rule antecedent opens up another possibility to express exceptions in moral rules. For instance, one can express an exception about lying, such as "If lying had been done to save an innocent from a murderer, then it would not have been wrong". That is, given a knowledge base about lying for human H:

$$lying\_wrong(H) \leftarrow lying(H), not\ make\_not(lying\_wrong(H)).$$

The antecedent of the above counterfactual can be represented as a rule:

$$make\_not(lying\_wrong(H)) \leftarrow save\_from\_murderer(H, I), innocent(I).$$

- Given that the conclusion of a counterfactual is some moral wrong W, abducing its antecedent in the form of intervention can be used for expressing a prevention of W, viz., "What could I have done to prevent a wrong W?".

While moral examples in this chapter are based on those from the literature with either their conceptual or empirical results readily available, a more systematic evaluation of computational moral reasoning should be considered, particularly when examples are taken from real world cases. For instance, an empirical study with human subjects, say in collaboration with cognitive scientists, may be conducted to provide insights about relevant counterfactuals in examining moral permissibility of the considered cases, as well as in examining argumentative processes of moral reasoning in justifying permissibility via counterfactuals. The study will then be able to guide the appropriate forms of counterfactual reasoning for representing morality issues.

References

1. Byrne, R.M.J.: The Rational Imagination: How People Create Alternatives to Reality. MIT Press, Cambridge (2007)
2. Collins, J., Hall, N., Paul, L.A. (eds.): Causation and Counterfactuals. MIT Press, Cambridge (2004)
3. Cushman, F., Young, L., Greene, J.D.: Multi-system moral psychology. In: Doris, J.M. (ed.) The Moral Psychology Handbook. Oxford University Press, Oxford (2010)
4. Foot, P.: The problem of abortion and the doctrine of double effect. Oxf. Rev. **5**, 5–15 (1967)
5. Greene, J.D., Nystrom, L.E., Engell, A.D., Darley, J.M., Cohen, J.D.: The neural bases of cognitive conflict and control in moral judgment. Neuron **44**, 389–400 (2004)
6. Han, T.A.: Evolution prospection with intention recognition via computational logic. Master's thesis, Technische Universität Dresden and Universidade Nova de Lisboa (2009)
7. Han, T.A., Saptawijaya, A., Pereira, L.M.: Moral reasoning under uncertainty. In: Proceedings of the 18th International Conference on Logic for Programming, Artificial Intelligence and Reasoning (LPAR). LNCS, vol. 7180, pp. 212–227. Springer, New York (2012)
8. Hauser, M., Cushman, F., Young, L., Jin, R.K., Mikhail, J.: A dissociation between moral judgments and justifications. Mind Lang. **22**(1), 1–21 (2007)
9. Hauser, M.D.: Moral Minds: How Nature Designed Our Universal Sense of Right and Wrong. Little Brown, London (2007)
10. Hoerl, C., McCormack, T., Beck, S.R. (eds.): Understanding Counterfactuals, Understanding Causation: Issues in Philosophy and Psychology. Oxford University Press, Oxford (2011)
11. Kamm, F.M.: Intricate Ethics: Rights, Responsibilities, and Permissible Harm. Oxford University Press, Oxford (2006)
12. Kowalski, R., Sadri, F.: Abductive logic programming agents with destructive databases. Ann. Math. Artif. Intell. **62**(1), 129–158 (2011)
13. Lopes, G., Pereira, L.M.: Prospective storytelling agents. In: Proceedings of the 12th International Symposium Practical Aspects of Declarative Languages (PADL). LNCS, vol. 5937. Springer, New York (2010)
14. Lopes, G., Pereira, L.M.: Visual demo of Princess-saviour Robot (2010). http://centria.di.fct.unl.pt/~lmp/publications/slides/padl10/quick_moral_robot.avi
15. Markman, K.D., Gavanski, I., Sherman, S.J., McMullen, M.N.: The mental simulation of better and worse possible worlds. J. Exp. Soc. Psychol. **29**, 87–109 (1993)
16. McIntyre, A.: Doctrine of double effect. In: Zalta, E.N. (ed.) The Stanford Encyclopedia of Philosophy, Fall 2011 edn. Center for the Study of Language and Information, Stanford University (2011). http://plato.stanford.edu/archives/fall2011/entries/double-effect/
17. Mikhail, J.: Universal moral grammar: theory, evidence, and the future. Trends Cogn. Sci. **11**(4), 143–152 (2007)
18. Newman, J.O.: Quantifying the standard of proof beyond a reasonable doubt: a comment on three comments. Law Probab. Risk **5**(3–4), 267–269 (2006)
19. Pereira, L.M., Dell'Acqua, P., Pinto, A.M., Lopes, G.: Inspecting and preferring abductive models. In: Nakamatsu, K., Jain, L.C. (eds.) The Handbook on Reasoning-Based Intelligent Systems, pp. 243–274. World Scientific Publishers, Singapore (2013)
20. Scanlon, T.M.: Moral Dimensions: Permissibility, Meaning, Blame. Harvard University Press, Cambridge (2008)

Part II
The Collective Realm

Chapter 9
Modeling Collective Morality via Evolutionary Game Theory

We have been addressing problems in machine ethics dealt with by using computational techniques.

In the preceding chapters, our research has focused on Computational Logic, particularly Logic Programming, and its appropriateness to model morality, namely moral permissibility, its justification, and the dual-process of moral judgments regarding the realm of the individual.

Now, in the sections of this chapter, we address the collective realm computationally, using Evolutionary Game Theory in populations of individuals, to report on norms and morality emergence. These populations, to start with, are not equipped with much cognitive capability, and simply act from a predetermined set of actions. Our research has shown that the introduction of cognitive capabilities, such as intention recognition, commitment, apology, forgiveness, and revenge, separately and jointly, reinforce the emergence of cooperation in populations, comparatively to their absence. We then prospect future work concerned with adding guilt.

In particular, we show:

- how learning to recognize intentions and committing resolve cooperation dilemmas;
- the evaluation of two strategies in the emergence of cooperation in groups, viz., avoidance versus restriction;
- the role of apology in committed versus commitment-free repeated interactions;
- how apology and forgiveness evolve to resolve failures in cooperative agreements;
- the role that guilt may play to prime apology and forgiveness.

9.1 The Collective Realm of Machine Ethics

The mechanisms of emergence and evolution of cooperation in populations of abstract individuals, with diverse behavioral strategies in co-presence, have been undergoing mathematical study via Evolutionary Game Theory (EGT), inspired in

L.M. Pereira and A. Saptawijaya, *Programming Machine Ethics*,
Studies in Applied Philosophy, Epistemology and Rational Ethics 26,
DOI 10.1007/978-3-319-29354-7_9

part on Evolutionary Psychology (EP). Their systematic study resorts to simulation techniques, thus enabling the study of aforesaid mechanisms under a variety of conditions, parameters, and alternative virtual games. The theoretical and experimental results have continually been surprising, rewarding, and promising. For a background on EGT and its use by EP we refer to [42].

In recent work, one of us (Pereira and the co-authors in the mentioned references) has initiated the introduction, in such groups of individuals, of cognitive abilities inspired on techniques and theories of Artificial Intelligence, namely those pertaining to Intention Recognition, Commitment, Apology, Forgiveness and Revenge (separately and jointly), encompassing errors in decision-making and communication noise. As a result, both the emergence and stability of cooperation become reinforced comparatively to the absence of such cognitive abilities. This holds separately for Intention Recognition, for Commitment, for Apology, and for Forgiveness and Revenge, and even more so when they are jointly engaged.

The subsequent sections aim to sensitize the reader to these Evolutionary Game Theory based issues, results and prospects, which are accruing in importance for the modeling of minds with machines, with impact on our understanding of the evolution of mutual tolerance and cooperation, and of the arising of moral norms. Recognition of someone's intentions, which may include imagining the recognition others have of our own intentions, and may comprise not just some error tolerance, but also a penalty for unfulfilled commitment though allowing for apology, can lead to evolutionary stable win/win equilibriums within groups of individuals, and perhaps amongst groups. The recognition and the manifestation of intentions, plus the assumption of commitment—even whilst paying a cost for putting it in place—and the acceptance of apology, are all facilitators in that respect, each of them singly and, above all, in collusion.

9.2 Software Sans Emotions but with Ethical Discernment

In this section we address the emergence of cooperation both in the individual and collective realms, sans emotions but with ethical discernment. And we justify epistemologically why computer studies, programs and simulations can be used to understand such issues and afford us with insight into their human counterparts.

9.2.1 Introduction

Some of our previous research [26, 44–46, 49] has focused on using logic programming techniques to computational modeling of morality sans emotions. In the realm of the individual, we have addressed questions of permissibility and the dual-process of moral judgments by framing together ingredients that are essential to moral agency: abduction, integrity constraints, preferences, argumentation, counterfactuals, and updates. Computation over these ingredients has become our vehicle for modeling the dynamics of moral cognition within a single agent, without addressing the cultural dimension, because this is still absent in machines. In the collective realm, we have reported on computational moral emergence [23], again sans emotions, using techniques from Evolutionary Game Theory (EGT). We have shown that the introduction of cognitive abilities, like intention recognition, commitment, revenge, apology, and forgiveness, reinforce the emergence of cooperation in diverse populations, comparatively to their absence, by way of EGT models.

In studies of human morality, these distinct but interconnected realms—one stressing above all individual cognition, deliberation, and behavior; the other stressing collective morals and how they have emerged with evolution—seem separate but are synchronously evinced [46]. Our account affords plenty of room for an evolutionary phylogenetic emergence of morality, as illustrated below, thereby supplementing the limitations of focusing just on ontogeny. The bridging issues concern individual cognitive abilities and their deployment in the population. Namely the one of recognizing the intention of another, even taking into account how others recognize our intention; the abilities of requesting commitment, and of accepting or declining to commit; those of cooperating or defecting; plus those of apologizing, be it fostered by guilt, and of taking revenge or forgiving.

This section relies mainly on our collective realm research, and considers the modeling of distinct co-present strategies of cooperative and uncooperative behavior. Such driving strategies are associated with moral "emotions" that motivate moral discernment and substantiate ethical norms, leading to improved general conviviality on occasion, or not. To wit, we can model moral agency without explicitly representing embodied emotions, as we know them. Rather, such software-instantiated "emotions" are modeled as (un)conscious heuristics empowered in complex evolutionary games.

In this chapter we build on the background support of Chap. 1, since in the ground breaking work of Alan Turing extolled there, functionalism is employed to scaffold a philosophical perspective on emotions and morality. The next four subsections review materials from our EGT-based research in support of this perspective. This work has substantiated the philosophical viewpoint through an admixture of intention recognition, commitment, revenge, apology, and forgiveness. The final subsection conjectures on guilt, and its relationship with counterfactual reasoning, as a next natural step in our research program.

9.2.2 Learning to Recognize Intentions and Committing Resolve Cooperation Dilemmas

Few problems have motivated the amalgamation of so many seemingly unrelated research fields as has the evolution of cooperation [39, 51]. Several mechanisms have been identified as catalysers of cooperative behavior (see survey in [39, 51]).

Yet these studies, mostly grounded on evolutionary dynamics and game theory, have neglected the important role, which is played by intention recognition [16] in behavioral evolution. In our work [18, 19], we explicitly studied the role of intention recognition in the evolution of cooperative behavior. The results indicate that intention recognizers prevail against the most successful strategies in the context of the iterated Prisoner's Dilemma (e.g., win-stay-lose-shift, and tit-for-tat like strategies), and promote a significantly high level of cooperation, even in the presence of noise plus the reduction of fitness associated with the cognitive costs of performing intention recognition. Thus, our approach offers new insights into the complexity of—as well as enhanced appreciation for the elegance of—behavioral evolution when driven by elementary forms of cognition and learning ability.

Moreover, our recent research [23, 25] into the synergy between intention recognition and cooperative commitment sheds new light on promoting cooperative behavior. This work employs EGT methods in agent-based computer simulations to investigate mechanisms that underpin cooperation in differently composed societies. High levels of cooperation can be achieved if reliable agreements can be arranged. Formal commitments, such as contracts, promote cooperative social behavior if they can be sufficiently enforced, and the costs and time to arrange them provide mutual benefit. On the other hand, an ability to assess intention in others has been demonstrated to play a role in promoting the emergence of co-operation.

An ability to assess the intentions of others based on experience and observations facilitates cooperative behavior without resort to formal commitments like contracts. Our research found that the synergy between intention recognition and commitment strongly depends on the confidence and accuracy of the intention recognition. To reach high levels of cooperation, commitments may be unavoidable if intentions cannot be assessed with sufficient confidence and accuracy. Otherwise, it is advantageous to wield intention recognition to avoid arranging costly commitments.

Now, conventional wisdom suggests that clear agreements need to be made prior to any collaborative effort in order to avoid potential frustrations for the participants. We have shown [22] that this behavior may actually have been shaped by natural selection. This research demonstrates that reaching prior explicit agreement about the consequences of not honoring a deal provide a more effective road to facilitating cooperation than simply punishing bad behavior after the fact, even when there is a cost associated to setting up the agreement. Typically, when starting a new project in collaboration with someone else, it pays to establish up-front how strongly your partner is prepared to commit to it. To ascertain the commitment level one can ask for a pledge and stipulate precisely what will happen if the deal is not honored.

In our study, EGT is used to show that when the cost of arranging commitments (for example, to hire a lawyer to make a contract) is justified with respect to the benefit of the joint endeavor (for instance buying a house), and when the compensation is set sufficiently high, commitment proposers become prevalent, leading to a significant level of cooperation. Commitment proposers can get rid of fake co-operators that agree to cooperate with them yet act differently, also avoiding interaction with the bad guys that only aim to exploit the efforts of the cooperative ones.

But what happens if the cost of arranging the commitments is too high compared to the benefit of cooperation? Would you make a legal contract for sharing a cake? Our results show that in that case those that free ride on the investment of others will "immorally" and inevitably benefit. Establishing costly agreements only makes sense for specific kinds of projects. Our study shows that insisting that your partner share in the cost of setting up a deal leads to even higher levels of cooperation, suggesting the evolution of cooperation for a larger range of arrangement costs and compensations. This makes sense, as equal investment will ensure the credibility of the pledge by both partners. Agreements based on shared costs result in better friends.

We also compared this behavior with costly punishment, a strategy that does not make any prior agreements and simply punishes afterwards. Previous studies show that by punishing strongly enough bad behavior cooperation can be promoted in a population of self-interested individuals [10]. Yet these studies also show that the punishment must sometimes be quite excessive in order to obtain significant levels of cooperation. Our study shows that arranging prior agreements can significantly reduce the impact-to-cost ratio of punishment. Higher levels of cooperation can be attained through lower levels of punishment. Good agreements make good friends indeed.

9.2.3 Emergence of Cooperation in Groups: Avoidance Versus Restriction

Public goods, like food sharing and social health systems, may prosper when prior agreements to contribute are feasible and all participants commit to do so. Yet, free riders may exploit such agreements [22], thus requiring committers to decide not to enact the public good when others are not attracted to committing. This decision removes all benefits from free riders (non-contributors), but also from those who are wishing to establish the beneficial resource. In [17] we show, in the framework of the one-shot Public Goods Game (PGG) and EGT, that implementing measures to delimit benefits to "immoral" free-riders, often leads to more favorable societal outcomes, especially in larger groups and in highly beneficial public goods situations, even if doing so incurs in new costs.

PGG is the standard framework for studying emergence of cooperation within group interaction settings [51]. In a PGG, players meet in groups of a fixed size, and all players can choose whether to cooperate and contribute to the public good or to defect without contributing to it. The total contribution is multiplied by a constant factor and is then equally distributed among all. Hence, contributors always gain less than free riders, disincentivizing cooperation. In this scenario, arranging a prior commitment or agreement is an essential ingredient in motivating cooperative behavior, as abundantly observed both in the natural world [36] and lab experiments [6]. Prior agreements help clarify the intentions and preferences of other players [19].

Refusing agreements may be conceived as intending or preferring not to cooperate (the non-committers).

In [17], we extend the PGG to examine commitment-based strategies within group interactions. Prior to playing the PGG, commitment-proposing players ask their co-players to commit to contribute to the PGG, paying a personal proposer?s cost to establish that agreement. If all of the requested co-players accept the commitment, the proposers assume everyone will contribute. Those who commit yet later do not contribute must compensate the proposers [22]. As commitment proposers may encounter non-committers, they require strategies to deal with these individuals. Simplest is to not participate in the creation of the common good. Yet, this avoidance strategy, AVOID, also removes benefits for those wishing to establish the public good, creating a moral dilemma. Alternatively, one can establish boundaries on the common good, so that only those who have truly committed have (better) access, or so that the benefit of non-contributors becomes reduced. This is the RESTRICT strategy.

Our results lead to two main conclusions: (i) Both strategies can promote the emergence of cooperation in the one-shot PGG whenever the cost of arranging commitment is justified with respect to the benefit of cooperation, thus generalizing results from pairwise interactions [22]; (ii) RESTRICT, rather than AVOID, leads to more favorable societal outcomes in terms of contribution level, especially when group size and/or the benefit of the PGG increase, even if the cost of restricting is quite large.

9.2.4 Why Is It so Hard to Say Sorry?

When making a mistake, individuals are willing to apologize to secure further cooperation, even if the apology is costly. Similarly, individuals arrange commitments to guarantee that an action such as a cooperative one is in the others' best interest, and thus will be carried out to avoid eventual penalties for commitment failure. Hence, both apology and commitment should go side by side in behavioral evolution. In [24], we studied the relevance of a combination of these two strategies in the context of the Iterated Prisoner's Dilemma (IPD). We show that apologizing acts are rare in non-committed interactions, especially whenever cooperation is very costly, and that arranging prior commitments can considerably increase the frequency of apologizing behavior. In addition we show that, with or without commitments, apology resolves conflicts only if it is sincere, i.e. costly enough. Most interestingly, our model predicts that individuals tend to use a much costlier apology in committed relationships than otherwise, because it helps better identify free riders, such as fake committers.

Apology is perhaps the most powerful and ubiquitous mechanism for conflict resolution [1, 40], especially among individuals involving in long-term repeated interactions (such as a marriage). An apology can resolve a conflict without having to involve external parties (e.g., teachers, parents, courts), which may cost all sides of the conflict significantly more. Evidence supporting the usefulness of apology abounds, ranging from medical error situations to seller-customer relationships [1]. Apology has been implemented in several computerized systems, such as human-computer interaction and online markets, to facilitate users' positive emotions and cooperation [58, 59].

The IPD has been the standard model to investigate conflict resolution and the problem of the evolution of cooperation in repeated interaction settings [3, 51]. The IPD game is usually known as a story of tit-for-tat (TFT), which won both Axelrod's tournaments [3]. TFT cooperates if the opponent cooperated in the previous round, and defects if the opponent defected. But if there can be erroneous moves due to noise (i.e. an intended move is wrongly performed), the performance of TFT declines, because an erroneous defection by one player leads to a sequence of unilateral cooperation and defection. A generous version of TFT, which sometimes cooperates even if the opponent defected [38], can deal with noise better, yet not thoroughly. For these TFT-like strategies, apology is modeled implicitly as one or more cooperative acts after a wrongful defection.

In [24], we describe a model containing strategies that explicitly apologize when making an error between rounds. An apologizing act consists in compensating the co-player an appropriate amount (the higher the more sincere), in order to ensure that this other player cooperates in the next actual round. As such, a population consisting of only apologizers can maintain perfect cooperation. However, other behaviors that exploit this apologetic behavior could emerge, such as those that accept apology compensation from others but do not apologize when making mistakes (fake apologizers), destroying any benefit of the apology behavior. Employing EGT [51], we show that when the apology occurs in a system where the players first ask for a commitment before engaging in the interaction [15, 20, 21], this exploitation can be avoided. Our results lead to these conclusions: (i) Apology alone is insufficient to achieve high levels of cooperation; (ii) Apology supported by prior commitment leads to significantly higher levels of cooperation; (iii) Apology needs to be sincere to function properly, whether in committed relationships or commitment-free ones (which is in accordance with existing experimental studies, e.g., in [40]); (iv) A much costlier apology tends to be used in committed relationships than in commitment-free ones, as it can help better identify free-riders such as fake apologizers: "*commitments bring about sincerity*".

In Artificial Intelligence and Computer Science, apology [58, 59] and commitment [60, 61] have been widely studied, namely how their mechanisms can be formalized, implemented, and used to enhance cooperation in human-computer interactions and online market systems [58, 59], as well as general multi-agent systems [60, 61]. Our study provides important insights for the design and deployment of such mechanisms; for instance, what kind of apology should be provided to customers when mistakes are made, and whether apology can be enhanced if complemented with commitments to ensure cooperation, e.g., compensation for customers who suffer wrongdoing.

9.2.5 Apology and Forgiveness Evolve to Resolve Failures in Cooperative Agreements

Making agreements on how to behave has been shown to be an evolutionarily viable strategy in one-shot social dilemmas. However, in many situations agreements aim to establish long-term mutually beneficial interactions. Our analytical and numerical results [33] reveal for the first time under which conditions revenge, apology and forgiveness can evolve, and deal with mistakes within on-going agreements in the context of the IPD. We showed that, when agreement fails, participants prefer to take revenge by defecting in the subsisting encounters. Incorporating costly apology and forgiveness reveals that, even when mistakes are frequent, there exists a sincerity threshold for which mistakes will not lead to the destruction of the agreement, inducing even higher levels of cooperation. In short, even when to err is human, revenge, apology and forgiveness are evolutionarily viable strategies, playing an important role in inducing cooperation in repeated dilemmas.

Using methods from EGT [27, 51], we provide analytical and numerical insight into the viability of commitment strategies in repeated social interactions, modeled through the IPD [2]. In order to study commitment strategies in the IPD, a number of behavioral complexities need to be addressed. First, agreements may end before the recurring interactions are finished. As such, strategies need to take into account how to behave when the agreement is present and when it is absent, on top of proposing, accepting or rejecting such agreements in the first place. Second, as shown within the context of direct reciprocity [57], individuals need to deal with mistakes made by an opponent or by themselves, caused for instance by "trembling hands" or "fuzzy minds" [39, 51]. A decision needs to be made on whether to continue the agreement, or end it collecting the compensation owed from the other's defection.

As errors might lead to misunderstandings or even breaking of commitments, individuals may have acquired sophisticated strategies to ensure that mistakes are not repeated or that profitable relationships may continue. Revenge and forgiveness may have evolved exactly to cope with those situations [7, 34]. The threat of revenge, through some punishment or withholding of a benefit, may discourage interpersonal harm. Yet, often one cannot distinguish with enough certainty if the other?s behavior is intentional or just accidental [12, 18]. In the latter case, forgiveness provides a restorative mechanism that ensures that beneficial relationships can still continue, notwithstanding the initial harm. An essential ingredient for forgiveness, analyzed in our work, seems to be (costly) apology [34], a point emphasized in [52].

The importance of apology and forgiveness for sustaining long-term relationships has been brought out in different experiments [1, 40, 41, 54]. Apology and forgiveness is of interest as they remove the interference of external institutions (which can be quite costly to all parties involved), in order to ensure cooperation.

Creating agreements and asking others to commit to them provides a basic behavioral mechanism present at all the levels of society, playing a key role in social interactions [6, 36, 53]. Our work reveals how, when moving to repeated games, the detrimental effect of having a large arrangement cost is moderated, for a subsisting commitment can play its role for several interactions. In these scenarios, the most successful individuals are those who propose commitments (and are willing to pay their cost) and, following the agreement, cooperate unless a mistake occurs. But if the commitment is broken then these individuals take revenge and defect in the remaining interactions, confirming analytically what has been argued in [7, 34]. This result is intriguing as revenge by withholding the benefit from the transgressor may lead to a more favorable outcome for cooperative behavior in the IPD, as opposed to the well-known reciprocal behavior such as TFT-like strategies. Forgivers only do better when the benefit-to-cost ratio is high enough.

Yet, as mistakes during any (long-term) relationship are practically inevitable, individuals need to decide whether it is worthwhile to end the agreement and collect the compensation when a mistake is made or whether it is better to forgive the co-player and continue the mutually beneficial agreement. To study this question, the commitment model was extended with an apology-forgiveness mechanism, where apology was defined either as an external or individual parameter in the model. In both cases, we have shown that forgiveness is effective if it takes place after receiving an apology from the co-players. However, to play a promoting role for cooperation, apology needs to be sincere, in other words, the amount offered in the apology has to be high enough (yet not too high), which is also corroborated by recent experimental psychology [35]. This extension to the commitment model produces even higher cooperation levels than in the revenge-based outcome. In the opposite case, fake committers that propose or accept a commitment with the intention taking advantage of the system (defecting and apologizing continuously) will dominate the population. In this situation, the introduction of the apology-forgiveness mechanism destroys the increase of the cooperation level that commitments by themselves produce. Thus, there is a lower-limit on how sincere apology needs to be, as below this limit apology and forgiveness even reduce the level of cooperation one could expect from simply taking revenge. It has been shown in previous works that mistakes can induce the outbreak of cheating or intolerant behavior in society [31, 32], and only a strict ethics can prevent them [32], which in our case would be understood as forgiving only when apology is sincere.

Commitments in repeated interaction settings may take the form of loyalty [4, 50], which is different from our commitments regarding posterior compensations, for we do not assume a partner choice mechanism. Loyalty commitment is based on the idea that individuals tend to stay with or select partners based on the length of their prior interactions. We go beyond these works by showing that, even without partner choice, commitment can foster cooperation and long-term relationships, especially when accompanied by sincere apology and forgiveness whenever mistakes are made.

Ohtsubo's experiment [40] shows that a costlier apology is better at communicating sincerity, and as a consequence will be more often forgiven. This observation is shown to be valid across cultures [54]. In another laboratory experiment [12], the authors showed apologies work because they can help reveal the intention behind a wrongdoer's preceding offense. In compliance with this observation, in our model, apology best serves those who intended to cooperate but defect by mistake.

Despite the fact that "to err is human" [47], our research results demonstrate that behaviors like revenge and forgiveness can evolve to cope with mistakes, even when they occur at high rates. Complicating matters is that mistakes are not necessarily intentional, and that even if they are then it might still be worthwhile to continue a mutually beneficial agreement. Here, a sincerity threshold exists whereby the cost of apologizing should exceed the cost of cooperating if the encouragement of cooperation is the goal.

9.2.6 Guilt for Non-humans

A natural future extension of our work on intention recognition, commitment, revenge, apology, and forgiveness involves adding guilt, shame, and confession with surplus apology. We argue next that the emotion of guilt, in the sense of actual harm done to others from inappropriate action or inaction, is worthwhile to incorporate in evolutionary game models, as it can lead to increased cooperation, whether by promoting apology or by inhibiting defection. The study thereof can then transpire to abstract and concrete populations of non-human agents.

9.2.6.1 Psychological Background

Theorists conceive of shame and guilt as belonging to the family of self-conscious emotions [13, 29, 55], invoked through self-reflection and self-evaluation. Though both have evolved to promote cooperation, guilt and shame can be treated separately. Guilt is an inward private phenomenon, though it can promote apology, and even spontaneous public confession. Shame is inherently public, though it may too lead to apology and to the request for forgiveness [52]. Shame, however, hinges on being caught, failing to deceive, and the existence of a reputation mechanism.

The philosopher Martin Buber [5] underlined the difference between the Freudian notion of guilt, based on internal conflicts, and *existential guilt*, based on actual harm done to others, which is the sense we are considering here. On transgression or error, the self renders judgment on itself. This self-evaluation may be explicit or implicit, conscious or unconscious. Shame and guilt typically arise from the blaming self-judgment about one's negative personal attributes (shame) or about negative harmful personal behavior or failure to act to prevent harm (guilt). Shame is often conducive to hiding and anger, and guilt is often conducive to admission and reparative action. Guilt is considered empathic whereas shame not [56]. We leave out shame for now, because it involves reputation, and concentrate on guilt instead.

To avoid or attempt to prevent blame assignment that might result from inappropriate action or inaction, there exists a guilt mechanism concerned not just with a posteriori guilt for a harm actually intended, but functions as well a priori, preventing harm by wishing to avoid guilt. A posteriori outward admission of guilt may serve to preempt punishment, whenever harm detection and blame become foreseeable.

We know too that guilt may be alleviated by private confession (namely to a priest or a psychotherapist) plus the renouncing of past failings in future. Because of their private character, such confessions and atonements, given their cost (prayers or fees), render temptation defecting less probable. Public or open confession of guilt can be coordinated with apology for better effect, and the cost appertained to some common good (like charity), or as individual compensation to injured parties.

More generally, Frank [14, pp. 46, 53] has suggested humans have been endowed during evolution with the means to solve problems of commitment by means of "moral sentiments", to wit, those of anger, contempt, disgust, envy, greed, shame, and guilt. Moral sentiments help solve such problems because honest manifestation of certain emotions make commitments more credible.

In particular, the promises of an agent known to be prone to guilt are therefore trustworthier. Heightened anger towards non-confessed guilt might be triggered by intention recognition, thereby putting pressure on guilt admission. Exhibition of guilt proneness by an agent may assuage other agents that defection by the agent was not intended, even when it might be clear it was not, since intention ascription by others is not perfect.

If intentions provide such a role in determining due apology, how can one read an offender's mental states? We regularly judge the mental states of others, and the notion of *mens rea*[1] in criminal law depends on this ability. An offender's emotions, namely feelings of guilt, provide a measure of his mental states [52]. Fake emotions better be discerned, of course. Smith [52] elaborates on detecting and distinguishing guilt, shame, embarrassment, remorse, and regret.

From Smith [52, pp. 101–103] we quote, abridging:

> According to [48], "shame is the emotion evoked by shocks to our self-respect" but we feel guilty when we "act [...] contrary to [our] sense of right and justice." Both involve our sense of morality, but in guilt "we focus on the infringement of just claims of others and the injury we have done to them, and on their probable resentment and indignation should they discover our deed."
>
> In shame "we feel struck by the loss to our self-esteem and our inability to carry out our aims: we sense the diminishment of self from our anxiety about the lesser respect that others may have for us and from our disappointment with ourselves for failing to live up to our ideals." A single wrongdoing might provoke feelings of both shame and guilt, but the primary distinction involves the emphasis on either my disappointment with myself (shame) or my concern for the victims and norms I have transgressed (guilt).
>
> [p. 102] Understood in accordance with the earlier description, guilt would seem like an appropriate emotional component of apology because it accompanies the recognition of wrongdoings as such. When we identify and share a commitment to the value underlying a transgression, guilt would appear to designate the corresponding emotion. As an undesirable emotion, guilt also spurs us to undertake the reform and redress likely to free us from its clutches.

---

[1] *Mens rea*, the intention or knowledge of wrongdoing that constitutes part of a crime, as opposed to the action or conduct of the accused. Compare with *actus reus*, action or conduct that is a constituent element of a crime, as opposed to the mental state of the accused. ORIGIN: mid 19th cent.: Latin, literally '*guilty mind*.'

> [p. 103] Negative emotions can have a deterrent value in that potential offenders may resist urges to commit offenses if they wish to avoid the unpleasant feelings of guilt or shame that may accompany their deed. Negative emotions may also serve rehabilitative objectives because an experience of guilt may move an offender to reform her behavior.

Furthermore, according to [56, pp. 494–496], guilt-prone individuals are inclined to manage anger constructively, and disinclined toward aggression. And they are less prone to defection and noise.

Guilt is the quintessential moral emotion: it promotes the acknowledgment of wrongdoing, the acceptance of responsibility, and the taking of reparative action. Expressions of guilt restore equanimity, reaffirm fairness, and compensate transgressions. Guilt leads to positive intra- and inter-personal processes, especially in contexts requiring cooperation.

9.2.6.2 Evolutionary Background on Guilt and Cooperation

If foreseen guilt prevents harm and absence of harm prevents possible retaliation and/or loss of reputation, then it would seem that a priori guilt would be evolutionarily advantageous. A posteriori guilt, on the other hand, would be evolutionarily advantageous because conducive to increased amount/possibility of apology, and we have seen apology is advantageous. Also apology reduces the pain of guilt.

Evolutionarily, guilt is envisaged as an in-built mechanism that tends to prevent wrong doing because of the internal self suffering it creates, and, should the wrong doing have taken place, it puts internal pressure on confession (admitting the wrong) and follow-up costly apology and penance, plus an expectation of forgiveness, so as to alleviate and dispel the internal suffering produced by guilt [11, 56]. Using the iterated Prisoner's Dilemma and Ultimatum games [28] found that guilt is a key factor in increasing cooperation among players.

It would pay for guilt to have spread once it appears, on two counts: as an inhibition mechanism and as a forgiveness stimulus. A third count, in our view, is that counterfactually thinking about guilt or shame is useful to prevent their future arising, a process of self-cleansing or self-debugging [37]. Counterfactual thinking, in turn, arises to explain causality in general, being subsequently used to handle guilt and self-improvement. Hence, guilt-dealing mechanisms seem to be evolutionarily advantageous for cooperation.

It seems clear that an evolutionary anthropological case about guilt has been made and accepted in the literature for intention recognition. Intentionality matters crucially to distinguish intended actions from noise, from accidental actions, and from side effects, since non-intended harms should be discounted because they are unavoidable, and a revenge arms race would be pernicious to all.

The notion of intentionality is ascribed to agents, and in particular to gods and nature spirits. The latter ones have the power to exercise justice with regard to acts of killing animals and slashing cultivars (which are living beings). When it doesn't rain, or some tragedy happens, it is ascribed to wrath of the gods or animal spirits symbolized by totems. Performing human or other sacrifices is meant to atone and apologize for the harm done to other living beings.

That's how guilt has arisen evolutionarily, as humans know that the spirits know what you did (and even about thoughts and memories inside your head about doing it on purpose) though no one might have witnessed your deed. Guilt and apology are primed by their moral disapproval and by revenge.

Population morality in turn arises for a great number of reasons (namely starting with mutualism and following on to contractualism), plus in particular making sure that one's intentions can be explained and do follow the accepted rules and exceptions. The literature on morality already makes the case in detail, albeit according to distinct schools of thought. Some of these are closer than others to the mechanisms that we know in AI.

The Baldwin effect describes an inclination for general learning mechanisms to open the way to domain-specific adaptations. Agent-based simulation models have shown that a moderate bias toward prosocial behavior is favored in evolving populations where punishment for anti-social behavior becomes dominant [8]. Guilt may be a primer for a Baldwin effect by which harm is partly avoided, and where punishment for harm is diminished by forgiveness of costly confession and apology. Moreover, the pain of guilt, even a non-confessed one, acts as internal punishment, and so may serve as an internal evaluation mechanism to undermine one's defection [9, 30]. Guilt signaling, having a cost in terms of accrued apology, comes under the "handicap principle" [62, 63], and that makes it "honest" or reliable, rather than deceitful.

Hence, on these counts, guilt will have been selected for prosociality.

9.2.6.3 Guilt Treating by Evolutionary Game Theory

The evolutionary issue about guilt is whether it is more worthwhile than the absence of guilt, with respect to emergence of cooperation. One would speak of guilty explicitly, and show that it's worthwhile, thus explaining its appearance on the evolutionary scene.

Guilt is widespread, for some reason, and we should show that it naturally connects with apology and forgiveness mechanisms, because of its emergent evolutionary advantage. Moreover, it does not seem too difficult to incorporate guilt into present frameworks involving apology and forgiveness [33]. It would mean duplicating each possibly defecting strategy into one experiencing guilt and a corresponding guiltless one.

This would open the way to the treatment of emotions as evolutionary mechanisms scaffolding cooperation, guilt being a widely acknowledged one. Furthermore, it would show that one does not need a specific kind of body (namely an anthropomorphic one) for guilt to be a functional useful emotion in population settings where cooperation is good value.

One may focus on emotions as being strategies in abstract evolutionary population games, sans specific embodiment [43]. One adds guilt (and, for that matter, possibly guilt promoted counterfactual reasoning) as an evolved means to trigger costly apology, to expect a better chance of forgiveness, and to assuage the inner guilt that prompts the triggering.

The hypothesis, then, is that the emergence of guilt in a population is evolutionarily advantageous.

We can test this hypothesis via our already existing model comprising apology, revenge, and forgiveness, by piggybacking guilt onto them [33]. We might introduce a zero/one guilt parameter, which, on defaulting, not only increases the probability of apology (confession), but also, spontaneously pays a costlier apology, as the means to atone for internal guilt (through the redressing towards the co-player), rather than simply apologizing.

On the other hand, the co-player will more readily accept a guilty apology and forgive. Indeed, this co-player attitude will favor in the population his own forgiveness by others, in case of his confession of guilt, instead of simple apology in the absence of guilt.

The prediction is that guilt will facilitate and speed-up the emergence of cooperation. In spite of initial its heavier cost, in time the cost will be recuperated within the guilt-ridden population. One reason being that it is compensated by the costlier guilt apology of others, another reason being that it is more conducive to forgiveness.

To emphasize our point: the experiments would need to explore different initial situations from the ones already considered. Instead of albeit different but homogeneous values of costly apology, the population would start with heterogeneous values of overall costly apology: a base cost (the apology) plus the added guilt cost.

Testing would mean that in the social imitation step the guilt and the forgiveness thresholds would also be copied, not just the strategy. One would start with a good mixture in the population of these two threshold factors, including zero guilt, within each of the strategies that defect and also within those that forgive, to see which factors pervade, for some apology compensation.

So, for each defaulting strategy, there would exist in the population both individuals with guilt and those without (say 50–50 % at the start). For the moment we would fix the forgiveness threshold. Guilt or its absence would be transmitted by social imitation too.

The base hypothesis is that when there is guilt in the starting population then the most frequent stationary distribution includes guilt and better cooperation. For which level of guilt this would happen would have to be experimentally found. Next, for that best level of guilt, we want to find a best level of probability of forgiveness, still starting with a mixed guilty/non-guilty population.

An additional possibility is to investigate guilt as a mechanism that diminishes defection. This would probably be tied to intention recognition, since guilt will have evolved as a fear about the detection of harm done (see above).

9.3 Concluding Remarks

To our knowledge, collective morality and its unavoidable entwining with individuals' cognitive moral abilities, which enact albeit distinct and possibly conflicting moral norms, has not been addressed in machine ethics research and its applications. We hope to have set forth the case for giving attention to the importance and richness of bridging the two realms in future research agendas, an issue into which we delve further in the next chapter.

References

1. Abeler, J., Calaki, J., Andree, K., Basek, C.: The power of apology. Econ. Lett. **107**(2), 233–235 (2010)
2. Axelrod, R., Hamilton, W.D.: The evolution of cooperation. Science **211**, 1390–1396 (1981)
3. Axelrod, R.: The Evolution of Cooperation. Basic Books, New York (1984)
4. Back, I., Flache, A.: The adaptive rationality of interpersonal commitment. Ration. Soc. **20**, 65–83 (2008)
5. Buber, M.: Guilt and guilt feelings. Psychiatry **20**(2), 114–129 (1957)
6. Cherry, T.L., McEvoy, D.M.: Enforcing compliance with enviromental agreements in the absence of strong institutions: an experimental analysis. Environ. Resour. Econ. **54**(1), 63–77 (2013)
7. Cullough, M.E., Kurzban, R., Tabak, B.A.: Evolved mechanisms for revenge and forgiveness. In: Shaver, P.R., Mikulincer, M. (eds.) Human Agression and Violence: Causes, Manisfestations, and Consequences. American Psychological Association, Washington (2011)
8. Cushman, F.A., Macendoe, O.: The coevolution of punishment and prosociality among learning agents. In: Proceedings of 31st Annual Conference of the Cognitive Science Society. Cognitive Science Society, Austin, TX (2009)
9. Damasio, A.: Descartes's Error. Avon, New York (1994)
10. Fehr, E., Gachter, S.: Altruistic punishment in humans. Nature **415**, 137–140 (2002)
11. Fessler, D.M.T., Haley, K.J.: The strategy of affect: emotions in human cooperation. Genetic and Cultural Evolution of Cooperation. The MIT Press, Cambridge (2012)
12. Fischbacher, U., Utikal, V.: On the acceptance of apologies. Games Econ. Behav. **82**, 592–608 (2013)
13. Fischer, K.W., Tangney, J.P.: Self-conscious emotions and the affect revolution: framewok and introduction. In: Fischer, K.W., Tangney, J.P. (eds.) The Self-conscious Emotions: Shame, Guilt, Embarrassment, and Pride. Guilford Press, New York (1995)
14. Frank, R.: Passions within Reason: The strategic Role of the Emotions. W. W. Norton, New York (1988)
15. Han, T.A.: Intention Recognition, Commitments and Their Roles in the Evolution of Cooperation: From Artificial Intelligence Techniques to Evolutionary Game Theory Models. SAPERE. Springer, Berlin (2013)
16. Han, T.A., Pereira, L.M.: State-of-the-art of intention recognition and its use in decision making. AI Commun. **26**(2), 237–246 (2013)
17. Han, T.A., Pereira, L.M., Lenaerts, T.: Avoiding or restricting defectors in public goods games? J. R. Soc. Interface **12**(103) (2015)
18. Han, T.A., Pereira, L.M., Santos, F.C.: Intention recognition promotes the emergence of cooperation. Adapt. Behav. **19**, 264–279 (2011)
19. Han, T.A., Pereira, L.M., Santos, F.C.: Corpus-based intention recognition in cooperation dilemmas. Artif. Life **18**(4), 365–383 (2012)

20. Han, T.A., Pereira, L.M., Santos, F.C.: The emergence of commitments and cooperation. In: Proceedings of 11th International Conference on Autonomous Agents and Multiagent Systems (2012)
21. Han, T.A., Pereira, L.M., Santos, F.C.: Intention recognition, commitment, and the evolution of cooperation. In: Proceedings of IEEE Congress on Evolutionary Computation (2012)
22. Han, T.A., Pereira, L.M., Santos, F.C., Lenaerts, T.: Good agreements make good friends. Nat. Sci. Rep. **3**(2695) (2013). doi:10.1038/srep02,695
23. Han, T.A., Pereira, L.M., Santos, F.C., Lenaerts, T.: Emergence of cooperation via intention recognition, commitment, and apology—a research summary. AI Commun. **28** (2015). doi:10.3233/AIC-150672
24. Han, T.A., Pereira, L.M., Santos, F.C., Lenearts, T.: Why is it so hard to say sorry: the evolution of apology with commitments in the iterated prisoner's dilemma. In: IJCAI 2013, pp. 177–183. AAAI Press (2013)
25. Han, T.A., Santos, F.C., Lenaerts, T., Pereira, L.M.: Synergy between intention recognition and commitments in cooperation dilemmas. Nat. Sci. Rep. **5**(9312) (2015)
26. Han, T.A., Saptawijaya, A., Pereira, L.M.: Moral reasoning under uncertainty. In: Proceedings of 18th International Conference on Logic for Programming, Artificial Intelligence and Reasoning (LPAR), LNCS, vol. 7180, pp. 212–227, Springer (2012)
27. Hofbauer, J., Sigmund, K.: Evolutionary Games and Population Dynamics. Cambridge University Press, New York (1998)
28. Ketelaar, T., Au, W.T.: The effect of feeling guilt on the behaviour of uncooperative individuals in repeated social bargaining games: an affect as information interpretation of the role of emotion in social interaction. Cogn. Emot. **17**(3) (2003)
29. Lewis, M.: Thinking and feeling—the elephant's tail. In: Maher, C.A., Schwebel, M., Fagley, N.S. (eds.) Thinking and Problem Solving in the Developmental Process: International Perspectives. Erlbaum, NJ (1990)
30. Mameli, M.: The role of emotions in ecological and practical rationality. Emotion Evolution and Rationality. Oxford University Press, Oxford (2004)
31. Martinez-Vaquero, L.A., Cuesta, J.A.: Evolutionary stability and resistance to cheating in an indirect reciprocity model based on reputation. Phys. Rev. E **87**(5), 052810 (2013)
32. Martinez-Vaquero, L.A., Cuesta, J.A.: Spreading of intolerance under economic stress: results from a reputation-based model. Phys. Rev. E **90**(2), 022805 (2014)
33. Martinez-Vaquero, L.A., Han, T.A., Pereira, L.M., Lenaerts, T.: Apology and forgiveness evolve to resolve failures in cooperative agreements. Nat. Sci. Rep. **5**(10639) (2015)
34. McCullough, M.E.: Beyond Revenge, the Evolution of the Forgiveness Instinct. Jossey-Bass, San Fransisco (2008)
35. McCullough, M.E., Pedersen, E.J., Tabak, B.A., Carter, E.C.: Conciliatory gestures promote forgiveness and reduce anger in humans. Proc. Natl. Acad. Sci. U.S.A **111**, 11211–11216 (2014)
36. Nesse, R.M.: Evolution and the Capacity for Commitment. Russell Sage, New York (2001)
37. Niedenthal, P.M., Tangney, J.P., Gavanski, I.: "If only I weren't" versus "If only I hadn't": distinguishing shame and guilt in counterfactual thinking. J. Pers. Soc. Psychol. **67**, 585–595 (1994)
38. Nowak, M.A., Sigmund, K.: Tit for tat in heterogeneous populations. Nature **355**, 250–253 (1992)
39. Nowak, M.A.: Five rules for the evolution of cooperation. Science **314**(5805), 1560–1563 (2006)
40. Ohtsubo, Y., Watanabe, E.: Do sincere apologies need to be costly? Test of a costly signaling model of apology. Evolution and Human Behavior **30**(2), 114–123 (2009)
41. Okamoto, K., Matsumura, S.: The evolution of punishment and apology: an iterated prisoner's dilemma model. Evol. Ecol. **14**, 703–720 (2000)
42. Pereira, L.M.: Evolutionary tolerance. In: Magnani, L., Ping, L. (eds.) Philosophy and Cognitive Science—Western & Eastern Studies. SAPERE, pp. 263–287. Springer, Berlin (2012)

43. Pereira, L.M.: Software sans emotions but with ethical discernment. In: Silva, S.G. (ed.) Morality and Emotion: (Un)conscious Journey to Being. Routledge, London (2016)
44. Pereira, L.M., Saptawijaya, A.: Modelling morality with prospective logic. In: Anderson, M., Anderson, S.L. (eds.) Machine Ethics, pp. 398–421. Cambridge University Press, New York (2011)
45. Pereira, L.M., Saptawijaya, A.: Abduction and beyond in logic programming with application to morality. Accepted at Frontiers of Abduction, a special issue of IfCoLog Journal of Logics and their Applications. Available from (preprint) http://goo.gl/yhmZzy (2015)
46. Pereira, L.M., Saptawijaya, A.: Bridging two realms of machine ethics. In: White, J.B., Searle, R. (eds.) Rethinking Machine Ethics in the Age of Ubiquitous Technology. IGI Global, Pennsylvania (2015)
47. Pope, A.: An Essay on Criticism, part II. Lewis, W. Russel Street, Covent Garden (1711)
48. Rawls, J.: A Theory of Justice. Harvard University Press, Cambridge (1971)
49. Saptawijaya, A., Pereira, L.M.: The potential of logic programming as a computational tool to model morality. In: Trappl, R. (ed.) A Construction Manual for Robots' Ethical Systems: Requirements, Methods, Implementations, Cognitive Technologies. Springer, New York (2015)
50. Schneider, F., Weber, R.A.: Long-term commitment and cooperation. Tech. Rep. working paper series, University of Zurich, Department of Economics (2013)
51. Sigmund, K.: The Calculus of Selfishness. Princeton University Press, Princeton (2010)
52. Smith, N.: I Was Wrong: the Meanings of Apologies, vol. 8. Cambridge University Press, New York (2008)
53. Sterelny, K.: The Evolved Apprentice. MIT Press, Cambridge (2012)
54. Takaku, S., Weiner, B., Ohbuchi, K.: A cross-cultural examination of the effects of apology and perspective taking on forgiveness. J. Lang. Soc. Psychol. **20**, 144–166 (2001)
55. Tangney, J.P., Dearing, R.: Shame and Guilt. Guilford Press, New York (2002)
56. Tangney, J.P., Stuewig, J., Malouf, E.T., Youman, K.: Communicative functions of shame and guilt. In: Sterelny, K., Joyce, R., Calcott, B., Fraser, B. (eds.) Cooperation and Evolution. The MIT Press, Cambridge (2013)
57. Trivers, R.L.: The evolution of reciprocal altruism. Q. Rev. Biol. **46**, 35–57 (1971)
58. Tzeng, J.Y.: Toward a more civilized design: studying the effect of computers that apologize. Int. J. Hum.-Comput. Stud. **61**(3), 319–345 (2004)
59. Utz, S., Matzat, U., Snijders, C.: On-line repotation systems: the effects of feedback comments and reactions on building and rebuilding trust in on-line auctions. Int. J. Electron. Commer. **13**(3), 95–118 (2009)
60. Winikoff, M.: Implementing commitment-based interactions. In: Proceedings of 6th International Conference on Autonomous Agents and Multiagent Systems (2007)
61. Wooldridge, M., Jennings, N.R.: The cooperative problem-solving process. J. Log. Comput. **9**(4), 563–592 (1999)
62. Zahavi, A.: Mate selection—a selection for a handicap. J. Theo. Biol. **53**(1), 205–214 (1975)
63. Zahavi, A.: The Handicap Principle: A Missing Piece of Darwin's Puzzle. Oxford University Press, Oxford (1977)

Chapter 10
Bridging Two Realms of Machine Ethics

In prior chapters we have addressed issues and topics of machine ethics programming, whether from the individual or from the collective viewpoints, which we dub "the two realms."

Bridging capabilities between the two realms, to wit, the individual and collective, helps understand the emergent ethical behavior of agents in groups, and implements them not just in simulations, but in the world of future robots and their swarms. With our co-authors, have staked footholds on either side of the two realms gap, and promoted their mutually beneficial bridging.

In studies of human morality, these distinct interconnected realms are evinced too: one stressing above all individual cognition, deliberation, and behavior; the other stressing collective morals, and how they emerged. Of course, the two realms are necessarily intertwined, for cognizant individuals form populations, and the twain evolved jointly to cohere into collective norms, and into individual interaction. Evolutionary Biology, Evolutionary Anthropology and the Cognitive Sciences provide inspirational teachings to that effect.

The chapter is naturally organized as follows. First, on the basis of preceding chapters, we consider the bridging of these two realms in machine ethics. Last but not least, we ponder over the teachings of human moral evolution in this regard. A final coda foretells a road to be tread, and portends about ethical machines and us.

10.1 Bridging the Realms

We have examined in prior chapters two types of incursions, one into the individual's success in a fixed group, and the second into the evolving population realms of morality.

The first type resorts to individual cognition and reasoning to enable such individuals to successfully compete amongst free riders and deceivers. Such successful competition can be achieved by learning past interactions with them or by

L.M. Pereira and A. Saptawijaya, *Programming Machine Ethics*,
Studies in Applied Philosophy, Epistemology and Rational Ethics 26,
DOI 10.1007/978-3-319-29354-7_10

recognizing their intentions [16]. The second type emphasizes instead the emergence, in a population, of evolutionarily stable moral norms, of fair and just cooperation, that ably discard free riders and deceivers, to the advantage of the whole evolved population.

To this latter end, some cognitive abilities such as intention recognition, commitment, apology, forgiveness, revenge, and guilt (pending) were employed, singly or jointly, by instilling them into just some individual agents, which then become predominant and lastly invade the evolving population, whether in the context of pairwise interactions or of public good situations.

A fundamental question then arises, concerning the study of individual cognition in groups of often morally interacting multi-agents (that can choose to defect or cooperate with others), whether from such study we can obtain results equally applicable to the evolution of populations of such agents. And vice-versa, whether the results obtained in the study of populations carry over to groups of frequently interacting multi-agents, and under what conditions. Some initial Evolutionary Game Theory results into certain learning methods have identified a broad class of situations where this is the case [5, 17, 19]. A premium outstanding issue remains in regard to which cognitive abilities and circumstances the result may obtain in general, and for sure that will be the object of much new and forthcoming programs of research.

Specifically with respect to human morality, the answer to the above-enounced fundamental question would appear to be a resounding 'Yes'. For one, morality concerns both groups and populations, requires cognition, and will have had to evolve in a nature/nurture or gene/culture intertwining and reinforcement. For another, evolutionary anthropology, psychology, and neurology have been producing ever more consilient views on the evolution of human morality.

Their scientific theories and results must per force be kept in mind, and serve as inspiration, when thinking and rethinking about machine ethics. And all the more so because the machines will need to be ethical amongst us human beings, not just among themselves.

Presently, machine ethics is becoming an ever more pressing concern, as machines become ever more sophisticated, autonomous, and act in groups, among populations of other machines and of humans. Ethics and jurisprudence, and hence legislation, are however lagging much behind in adumbrating the new ethical issues arising from these circumstances.

On the other hand, the very study of ethics, and the evolution of human morality too, can now avail themselves of the experimental, computation theoretic, and robotic means to enact and simulate individual or group moral reasoning, in a plethora of circumstances. Likewise for the emergence of moral rules and behaviors in evolving populations.

Hence, having addressed in previous chapters the two computational types of models, in the next section we stress this double outlook, by bringing to the fore congenial present views and research on the evolution of human morality, hoping to reinforce the bridging ideas and paradigm we set forth.

Moreover, we take for granted that computational and robotic models can actually provide abstract and concrete insight on emerged human moral reality, irrespective of the distinct embodiments of man and machine.

10.2 Evolutionary Teachings

Added dependency on cooperation makes it more competitive to cooperate well. Thus, it is advantageous to invest on shared morals in order to attract partners who will partake of mutual and balanced advantages.

This evolutionary hypothesis inspired by mutualism [2]—itself a form of contractualism [1]—contrasts with a number of naturalist theories of morality, which make short shrift of the importance of cognition for cooperation. For example, the theory of reciprocity, in ignoring a wider cognitive capacity to choose and attract one's partners, forbids itself from explaining evolution on the basis of a cooperation market.

Indeed, when assigning all importance to population evolutionary mechanisms, naturalist theories tend to forget the evolution of cognition in individuals. Such theories habitually start off from evolutionary mechanisms for understanding the specificity of human morals: punishment [7, 20], culture [14, 20], political alliances [3, 10]. According to Baumard's hypothesis, morality does not emerge because humans avail themselves of new means for punishing free-riders or for recompensing cooperators, but simply because mutual help—and hence the need to find partners—becomes much more important.

In summary, it's the development of cooperation that induces the emergence of morals, and not the stabilization of morals (via punishment or culture) that promotes the development of cooperation.

Experimental results are in line with the hypothesis that the perfecting of human intuitive psychology is responsible for the emergence of morality, on the basis of an improved understanding of the mental states of others. This permits to communicate, not just to coordinate with them, and thus extend the domain cooperation, thereby leading to a disposition toward moral behaviors. For a systematic and thorough account of research into the evolutionary origins of morality, see [6, 15].

At the end of the day, one may consider three theories bearing on three different aspects of morality: the evaluation of interests for utilitarianism, the proper balance of interests for mutualism, and the discharging of obligations for the virtues principled.

A naturalistic approach to moral sense does not make the psychological level disappear to the benefit of the evolutionary one. To each its explanation level: psychology accounts for the workings of the moral sense; sociology, for the social context that activates it; and a cupola theory, for the evolution of causes that occasioned it [21]. Moral capability is therefore a "mechanism" amongst others [9], as are the concern for reputation, the weakness of the will, the power to reason, etc.

An approach that is at once naturalist and mutualist allows escape from these apparently opposite viewpoints: the psychological and the societal. At the level of psychological motivations, moral behavior does neither stem from egotism nor altruism. To the contrary, it aims at the mutual respect for everyone's attending interests. And, simultaneously, it obeys the logic of equity. At the evolutionary level, moral behavior is not contradictory with egotism because, in human society, it is often in our own interest to respect the interests of others. Through moral motivations, we avail ourselves of a means to reconcile the diverse individual interests. Morality vies precisely at harmonizing individual interest with the need to associate, and profit from cooperation, by adopting a logic of fairness.

The mutualist solution is not new. Contractualist philosophers have upheld it for some time. Notably, they have furnished detailed descriptions of our moral capacity [18, 22]. However, they never were able to explain why humans are enabled with that particular capacity: Why do our judgments seek equity? Why do we behave morally at all?

Without an explanation, the mutualist theory seems improbable: Why behave we as if an actual contract had been committed to, when in all evidence one was not?

Past and ongoing evolutionary studies, intertwining and bridging cognitive and population aspects, and both becoming supported on computational simulations, will help us find answers to that. In the process, rethinking machine ethics and its implementations.

According to [4], conscience and morality evolved, in the biological sense. Conscience evolved for reasons having to do with environments humans had to cope with prehistorically, and their growing ability to use group punishment to better their social and subsistence lives and create more equalized societies. His general evolutionary hypothesis is that morality began with having a conscience and that conscience evolution began with systematic but initially non-moralistic social control by groups.

This entailed punishment of individual "deviants" by bands of well-armed large-game hunters, and, like the ensuing preaching in favor of generosity, such punishment amounted to "social selection", since the social preferences of members and of groups as a whole had systematic effects on gene pools.

This punitive side of social selection adumbrates an immediate kind of "purpose", of large-brained humans actively and insightfully seeking positive social goals or avoiding social disasters arising out of conflict. No surprise the genetic consequences, even if unintended, move towards fewer tendencies for social predation and more towards social cooperation. Hence, group punishment can improve the quality of social life, and over the generations gradually shape the genotype in a similar direction.

Boehm's idea is that prehistoric humans made use of social control intensively, so that individuals who were better at inhibiting their own antisocial tendencies, by fear of punishment or by absorbing and identifying with group's rules, garnered a superior fitness. In learning to internalize rules, humankind acquired a conscience. At the beginning this stemmed from punitive social selection, having also the strong effect of suppressing free riders. A newly moralistic type of free-rider suppression helped evolve a remarkable capacity for extra-familial social generosity. That conscience gave us a primitive sense of right and wrong, which evolved the remarkable "empathy" which we are infused with today. It is a conscience that seems to be as much a Machiavellian risk calculator as a moral force that maximizes prosocial behavior, with others' interests and equity in mind, and minimizes deviance too. It is clear that "biology" and "culture" work together to render us adaptively moral.

Boehm believes the issue of selfish free riders requires further critical thought, and that selfish intimidators are a seriously neglected type of free rider. There has been too much of a single-minded focus on cheating dominating free rider theorizing. In fact, he ascertains us the more potent free riders have been alpha-type bullies, who simply take what they want. It is here his work on the evolution of hunter-gatherer egalitarianism enters, namely with its emphasis on the active and potentially quite violent policing of alpha-male social predators by their own band-level communities. Though there's a large literature on cheaters and their detection, free-rider suppression in regard to bullies has not been taken into account so far in the mathematical models that study altruism.

"For moral evolution to have been set in motion," Boehm [4] goes on, "more was needed than a preexisting capacity for cultural transmission. It would have helped if there were already in place a good capacity to strategize about social behavior and to calculate how to act appropriately in social situations."

In humans, the individual understanding that there exists a self in relation to others makes possible participation in moral communities. Mere self-recognition is not sufficient for a moral being with fully developed conscience, but a sense of self is a necessary first step useful in gauging the reactions of others to one's behavior and to understand their intentions. And it is especially important to realize that one can become the center of attention of a hostile group, if one's actions offend seriously its moral sensibilities. The capacity to take on the perspective of others underlies not just the ability of individuals in communities to modify their behavior and follow group imposed rules, but it also permits people acting as groups to predict and cope insightfully with the behavior of "deviants."

Social selection reduced innate dispositions to bully or cheat, and kept our conscience in place by self-inhibiting antisocial behavior. A conscience delivers us a social mirror image. A substandard conscience may generate a substandard reputation and active punishment too. A conscience supplies not just inhibitions, but serves as an early warning system that helps prudent individuals from being sanctioned.

Boehm [4] wraps up: "When we bring in the conscience as a highly sophisticated means of channeling behavioral tendencies so that they are expressed efficiently in terms of fitness, scenarios change radically. From within the human psyche an evolutionary conscience provided the needed self-restraint, while externally it was group sanctioning that largely took care of the dominators and cheaters. Over time, human individuals with strong free-riding tendencies—but who exercised really efficient self-control—would not have lost fitness because these predatory tendencies were so well inhibited. And if they expressed their aggression in socially acceptable ways, this in fact would have aided their fitness. That is why both free-riding genes and altruistic genes could have remained well represented and coexisting in the same gene pool."

For sure, we conclude, evolutionary biology and anthropology, like the cognitive sciences too [8, 11–13, 23], have much to offer in view of rethinking machine ethics, evolutionary game theory simulations of computational morality to the rescue.

10.3 Concluding Remarks

We have argued that the study of the aforementioned issues has come of age and is ripe with research opportunities, having communicated some of the inroads we explored, and pointed to the more detailed published results of what we have achieved, with respect to intention recognition, commitment, and mutual tolerance through apology, forgiveness, revenge, and guilt, within the overarching Evolutionary Game Theory context.

In realm of the individual, computation is indeed vehicle for the study and teaching of morality, namely in its modeling of the dynamics of knowledge and cognition of agents. In the collective realm, norms and moral emergence have been studied computationally in populations of rather simple-minded agents. By bridging these realms, cognition affords improved emerged morals in populations of situated agents. At the end of the day, we will certainly wish ethical machines to be convivial with us.

References

1. Ashford, E., Mulgan, T.: Contractualism. In: E.N. Zalta (ed.) The Stanford Encyclopedia of Philosophy, Fall 2014 edn. Center for the Study of Language and Information, Stanford University (2014). http://plato.stanford.edu/archives/fall2012/entries/contractualism/
2. Baumard, N.: Comment nous sommes devenus moraux: Une histoire naturelle du bien et du mal. Odile Jacob, Paris (2010)
3. Boehm, C.: Hierarchy in the Forest: the Evolution of Egalitarian Behavior. Harvard University Press, Cambridge (1999)
4. Boehm, C.: Moral Origins: the Evolution of Virtue, Altruism, and Shame. Basic Books, New York (2012)
5. Börgers, T., Sarin, R.: Learning through reinforcement and replicator dynamics. J. Econ. Theory **77**(1), 1–14 (1997)
6. Bowles, S., Gintis, H.: A Cooperative Species: Human Reciprocity and Its Evolution. Princeton University Press, Princeton (2011)
7. Boyd, R., Richerson, P.: Punishment allows the evolution of cooperation (or anything else) in sizable groups. Ethol. Sociobiol. **13**(3), 171–195 (1992)
8. Churchland, P.: Braintrust: What Neuroscience Tells Us about Morality. Princeton University Press, Princeton (2011)
9. Elster, J.: A plea for mechanisms. Social Mechanisms: An Analytical Approach to Social Theory. Cambridge University Press, Cambridge (1998)
10. Erdal, D., Whiten, A., Boehm, C., Knauft, B.: On human egalitarianism: an evolutionary product of machiavellian status escalation? Curr. Anthropol. **35**(2), 175–183 (1994)
11. Gazzaniga, M.S.: The Ethical Brain: The Science of Our Moral Dilemmas. Harper Perennial, New York (2006)

12. Greene, J.: Moral Tribes: Emotion, Reason, and the Gap Between Us and Them. The Penguin Press HC, New York (2013)
13. Hauser, M.D.: Moral Minds: How Nature Designed Our Universal Sense of Right and Wrong. Little Brown, London (2007)
14. Henrich, J., Boyd, R.: Why people punish defectors: weak conformist transmission can stabilize costly enforcement of norms in cooperative dilemmas. J. Theor. Biol. **208**(1), 78–89 (2001)
15. Krebs, D.L.: The Origins of Morality—An Evolutionary Account. Oxford University Press, Oxford (2011)
16. Pereira, L.M., Han, T.A.: Intention recognition with evolution prospection and causal bayesian networks. In: Madureira, A., Ferreira, J., Vale, Z. (eds.) Computational Intelligence for Engineering Systems: Emergent Applications. Intelligent Systems, Control and Automation: Science and Engineering Book Series, pp. 1–33. Springer, Dordrecht (2011)
17. Pinheiro, F.L., Pacheco, J.M., Santos, F.C.: From local to global dilemmas in social networks. PLoS ONE **7**(2), e32114 (2012). doi:10.1371/journal.pone.0032,114
18. Rawls, J.: A Theory of Justice. Harvard University Press, Cambridge (1971)
19. Segbroeck, S.V., Jong, S.D., Nowé, A., Santos, F.C., Lenaerts, T.: Learning to coordinate in complex networks. Adap. Behav. **18**(5), 416–427 (2010)
20. Sober, E., Wilson, D.: Unto Others: The Evolution and Psychology of Unselfish Behavior. Harvard University Press, Cambridge (1998)
21. Sperber, D.: Individualisme méthodologique et cognitivisme. In: Cognition et sciences sociales. Presses Universitaires de France, Paris (1997)
22. Thomson, J.J.: A defense of abortion. Philos. Publ. Aff. **1**(1), 47–66 (1971)
23. Tomasello, M.: A Natural History of Human Thinking. Harvard University Press, Cambridge (2014)

Part III
Coda

Chapter 11
Conclusions and Further Work

This book discusses the two realms of machine ethics, a field that is now becoming a pressing concern and receiving wide attention due to its growing importance.

In the individual realm, we explore the appropriateness of LP-based reasoning features to machine ethics. Our starting point has been identifying moral facets—through interdisciplinary literature study—that are amenable, in our view, to computational modeling. We focus on three moral facets: moral permissibility (with respect to the Doctrines of Double and Triple Effect, and Scanlonian contractualism), the dual-process model in moral decision making, and counterfactual thinking in moral reasoning.

The book makes a number of original inroads that exhibit a proof of possibility to systematically represent and reason about a variety of issues from the chosen moral facets by means of moral examples taken off-the-shelf from the morality literature. This is accomplished via a combination of Logic Programming features, where abduction serves as the basic mechanism for examining moral decisions. Indeed, the potential and suitability of Logic Programming, and computational logic in general, for machine ethics is identified and discussed at length in [7], on the heels of our work.

While there have been approaches that provide implementations in LP systems, e.g., [1, 2, 6], to the best of our knowledge, this is the first attempt for various LP-based reasoning features being considered, individually and in combination, in the field of machine ethics. Moreover, our choice to model Scanlonian contractualism, the dual-process model, including our formulation of counterfactuals to address moral permissibility are also novel in this field.

Concerning the collective realm, we hope to have abundantly set forth the case for paying due attention to its importance and richness. It is patent that collective morality and its unavoidable entwining with individuals' cognitive moral abilities, enacting distinct and possibly conflicting moral norms and behaviors, has not been much addressed in machine ethics research and its applications. The bridging of the two realms in research agendas is by now unavoidable.

L.M. Pereira and A. Saptawijaya, *Programming Machine Ethics*,
Studies in Applied Philosophy, Epistemology and Rational Ethics 26,
DOI 10.1007/978-3-319-29354-7_11

Given the broad dimension of the topic, the contributions in the book touch solely on a dearth of morality issues. Nevertheless, it prepares and opens the way for additional research towards employing various features available in LP-based reasoning and EGT to machine ethics.

Research in machine ethics should not only be considered abstractly. The three systems considered in this book are just a start to provide an implementation as proof of concept and a testing ground for experimentation. Given the possibility to bridge Prolog with other languages, QUALM may be employed for developing morality-related applications, e.g., a visual interactive storytelling for teaching morality, or for equipping role playing games with ethical considerations.

For philosophers and psychologists, who are not familiar with LP, to also benefit from machine ethics, an appropriate interface for using QUALM may also be built. For instance, INTERPROLOG STUDIO [5] can be customized to deal with specific syntax used in the modeling.

We mention in the Preface that this book is not intended as a proposal for a machine readily incorporating ethics, but as a proof of concept that our understanding of the considered moral facets can in part be computationally modeled and implemented using a combination of LP-based reasoning features. A system with a considerably large number of features will undoubtedly increase the complexity of having such a more general system. In this book, each of the three systems (ACORDA, PROBABILISTIC EPA, and QUALM) focuses on a different combination of features. On the other hand, limiting the application to a specific domain may alternatively help in reducing the complexity of the required features. In terms of machine ethics applications, it may also constrain morality issues to tackle and domain-specific ethical principles to focus on.

We draw attention to some salient philosophical considerations underpinning and surging from our work.

Disembodiment

In Chap. 1 we emphasized disembodiment about any particular hardware to justify the use of computers and computation for the purpose of modeling, developing and testing machine ethics. Furthermore, we wish to emphasize down the use of a formal declarative semantics approach to machine ethics, inasmuch morality is contextual and culturally diverse, more in line with an experimental and procedural semantics that passes a suite of tests, what is called Psychometric AI, more an engineering rather than a philosophical point of view [4], as in this book. Moreover, the diversity of approaches and standpoints to human ethics, as well as the study of the evolutionary roots of morality [8], counsels against excessive straightjacket formalization.

Learning and Morality

Morality is to be applied in a wide scope of situations. Big-data analysis consists of searching for buried patterns that have some kind of predictive power. However, one can't base machine ethics on predicting how to do the right thing. Morality is not about statistics, but about being right or wrong according to rules and exceptions, in specific circumstances. Present day wide scope machine learning from big data does not come up with rules, and so cannot explain and justify actions, which is a must for ethical machine actions to be accepted. The community should be well aware that such present day learning is inadequate for general machine morality. Only small, circumscribed, well-defined domains have been susceptible to rule generation through machine learning. Rules are all important for moral explanation, justification and argumentation.

Two Realms and Consilience

We underscored the importance of connecting the individual cognitive realm of morality—with its facets of dual process, intention recognition, preferences, several forms of reasoning combination, including meta-reasoning, etc.—and the realm of evolving populations of individuals having diverse strategies, where moral norms may emerge. Such salient connections [3], leading to integrated exploratory models that bypass traditional divisions between "levels" or "domains", require a broad consilience of the sciences and the humanities [10], convening namely ethics, cultural anthropology, cognitive science, evolutionary biology, psychology, AI, computer science, philosophy, economics, among others [9].

References

1. Anderson, M., Anderson, S.L.: EthEl: Toward a principled ethical eldercare robot. In: Proceedings of the AAAI 2008 Fall Symposium on AI in Eldercare (2008)
2. Anderson, M., Anderson, S.L., Armen, C.: MedEthEx: a prototype medical ethics advisor. In: Proceedings of the 18th Innovative Applications of Artificial Intelligence Conference (IAAI 2006)
3. Boyer, P.: From studious irrelevancy to consilient knowledge: modes of scholarship and cultural anthropology. In: Slingerland, E., Collard, M. (eds.) Creating Consilience: Evolution, Cognitive Science, and the Humanities. Oxford University Press, New York (2012)
4. Bringsjord, S.: Psychometric artificial intelligence. J. Exp. Theor. Artif. Intell. **23**(3), 271–277 (2011)
5. Calejo, M.: InterProlog Studio. http://interprolog.com/interprolog-studio (2014)
6. Ganascia, J.G.: Modelling ethical rules of lying with answer set programming. Ethics Inf. Technol. **9**(1), 39–47 (2007)
7. Kowalski, R.: Computational Logic and Human Thinking: How to be Artificially Intelligent. Cambridge University Press, New York (2011)
8. Krebs, D.L.: The evolution of a sense of morality. In: Slingerland, E., Collard, M. (eds.) Creating Consilience: Evolution, Cognitive Science, and the Humanities. Oxford University Press, New York (2012)
9. Pereira, L.M.: Evolutionary psychology and the unity of sciences - towards an evolutionary epistemology. In: Pombo, O., Torres, J.M., Symons, J., Rahman, S. (eds.) Special Sciences and the Unity of Science, Logic, Epistemology, and the Unity of Science, vol. 24. Springer, Dordrecht (2012)
10. Slingerland, E., Collard, M.: Creating Consilience: Evolution, Cognitive Science, and the Humanities. Oxford University Press, Oxford (2012)

Index

L.M. Pereira and A. Saptawijaya, *Programming Machine Ethics*,
Studies in Applied Philosophy, Epistemology and Rational Ethics 26,
DOI 10.1007/978-3-319-29354-7

Kolibri 360 GmbH
Ferdinand-Jühlke-Straße 7
99095 Erfurt, Deutschland
produktsicherheit@kolibri360.de